SPACE HANDBOOK

Astronautics and Its Applications

ROBERT W. BUCHHEIM

AND THE STAFF OF THE RAND CORPORATION

Space Handbook: Astronautics and Its Applications was originally published by Random House/Modern Library in 1958/1959. This RAND edition reflects the original layout.

Library of Congress Cataloging-in-Publication Data

978-0-8330-4223-1

Cover Design by Peter Soriano

Published 2007 by the RAND Corporation
1776 Main Street, P.O. Box 2138, Santa Monica, CA 90407-2138
1200 South Hayes Street, Arlington, VA 22202-5050
4570 Fifth Avenue, Suite 600, Pittsburgh, PA 15213-2665
RAND URL: http://www.rand.org/
To order RAND documents or to obtain additional information, contact
Distribution Services: Telephone: (310) 451-7002;
Fax: (310) 451-6915; Email: order@rand.org

FOREWORD

This handbook was prepared by The RAND Corporation for the Select Committee on Astronautics and Space Exploration, House of Representatives.

In the words of the committee, "A real need has been felt for an authoritative study in lay terms which would set forth clearly the present and definitely foreseeable state of the art of space flight. The committee, after careful consideration of alternatives, requested The RAND Corporation of Santa Monica, Calif., to undertake such a study. Under contract with the United States Air Force, RAND scientists and engineers have been in the forefront of objective investigation of such problems since World War II. RAND's reputation for integrity and independence particularly commended this nonprofit organization to our attention.

"The report which follows . . . represents the most comprehensive unclassified study on the subject now available. The report is confined to technical and scientific analysis, avoiding expressions of opinion on policy and administrative matters. It studiously avoids borderline speculative judgments on the pace of future development."

In accepting the report, the Select Committee passed a special resolution commending The RAND Corporation.

Preparation of the report was under the direction of Robert W. Buchheim, now head of the Aero-Astronautics Department in the Engineering Division of The RAND Corporation.

The report was forwarded on December 29, 1958, to the Honorable John W. McCormack, Majority Leader, and Chairman, Select Committee on Astronautics and Space Exploration, by the Committee Director and Chief Counsel, George J. Feldman, and Assistant Director Charles Sheldon II. The present volume has been revised, corrected, and brought up to date in the light of developments during the first two months of 1959.

The work of preparing this report was undertaken by The RAND Corporation with its own funds in the public interest. The views expressed in this handbook are not necessarily those of the United States Air Force or of the committee or any of its members.

This report was prepared under the technical direction of Robert W. Buchheim, with the assistance of the following members of the Staff of THE RAND CORPORATION:

R. L. Bjork
S. T. Cohen
C. M. Crain
M. H. Davis
Robert A. Davis
Stephen H. Dole
T. I. Edwards
R. H. Frick
R. T. Gabler
T. B. Garber
Carl Gazley, Jr.
Joseph M. Goldsen
Martin Goldsmith
S. M. Greenfield
E. C. Heffern
George A. Hoffman
Arnold L. Horelick
John H. Huth
Lloyd E. Kaechele
Hilde K. Kallmann
Amrom H. Katz
W. W. Kellogg
F. J. Krieger
H. A. Lang
Eugene Levin
Hans A. Lieske
Milton A. Margolis
David J. Masson
William R. Micks
Frederick S. Pardee
Sidney Passman
Louis N. Rowell
Richard Schamberg
Frederick T. Smith
Myron C. Smith
Peter Swerling
E. H. Vestine
E. P. Williams
Albert G. Wilson

CONSULTANTS:

Jesse L. Greenstein, California Institute of Technology
Samuel Herrick, University of California, Los Angeles

EDITORS:

Neil J. Horgan
Malcolm A. Palmatier
J. Vogel

CONTENTS

The reader's attention is called to the fact that the illustration, table, and reference numbers used in this publication are complete within each section, rather than consecutive throughout the text.

LIST OF ILLUSTRATIONS

LIST OF TABLES

PART 1

INTRODUCTION

1

Introduction

A. Historical Notes

The early history of space flight is really the history of an idea deeply imbedded in the general stream of development of human thought about the nature of the universe. The notion of flight to the Moon followed almost instantaneously upon the arrival of the idea that the Moon might be another solid sphere akin to the Earth. The evolution of these speculations from ancient times is treated in a fascinating manner by Willy Ley, an acknowledged historian of astronautics.[1]

The first glimmer of a chance to convert fanciful notions of extraterrestrial flight into an idea with engineering significance came with the invention of the rocket.

The first applications of rocket propulsion were, with little doubt, military, and rockets have had a long and varied career in military service, mostly as on-and-off rivals of artillery.

The current feasibility of space activities is clearly the product of modern weapons developments, the first substantial step having been taken in the German V-2 program.[2] This beginning has been greatly extended in the intermediate range (IRBM) and intercontinental ballistic missile (ICBM) programs in the United States and the Soviet Union.

For most of their long history, military rockets were viewed as "gunless artillery" and estimates of their merits were based on comparisons of their performance with that of competing artillery pieces. Only rather recently have rockets been looked upon as devices applicable to a class of activity far removed from anything achievable by artillery projectors; they are now more nearly rivals or companions of long-range bomber aircraft.

It is interesting to note that Maj. J. R. Randolph, an officer of the United States Army Reserve, deduced 20 years ago that the rocket has two likely applications: as gunless artillery, and for bombardment over intercontinental ranges.[3] His assessment, with respect to the application now labeled "ICBM," was based on data developed by investigators interested in interplanetary flight.

Major Randolph also suggested the possibility of using large liquid-propellant rockets of ICBM class as boosters for manned intercontinental bombing vehicles—a notion now being implemented in the Dyna-Soar program (see p. 178). These ideas about rockets are a strikingly good broad outline of the more detailed program developed at Peenemünde for extension of German rocket development to intercontinental scope.[4, 5] There have been suggestions that this general theme has also been operative in U. S. S. R. development planning.[6]

Military and peaceful application of rockets was pioneered in the United States by Dr. Robert Goddard, who was in charge of War Department rocket work during World War I. Dr. Goddard advanced the state of the rocket art in many ways in the twenties and thirties, and saw the great potential for scientific experimentation inherent in rocket propulsion as a means of reaching altitudes otherwise unattainable.[7]

The principal early work in the technological field of space flight was done in Russia, Germany, and the United States. The chief United States effort was that of Goddard. The early German work was done by H. Oberth, beginning in the 1920's. Russian efforts commenced at a substantially earlier date, giving them a clear and valid claim to a "first." Russian activity began with the work of Meshcherskii and Tsiolkovskii near the end of the 19th century.[8, 9] Tsiolkovskii is generally recognized as the father of astronautics. Considerable work, both theoretical and experimental, was accomplished in the U. S. S. R. in the 1920's and 1930's.

Serious and substantial Government-sponsored rocket-research programs were established in Germany in about 1930, in the Soviet Union sometime in or before 1934, and in the United States in 1942. The astronautical activities of the

United States and the U. S. S. R. will be discussed in greater detail below.

B. General Nature of Astronautics

Even in its present early and uncertain state, astronautics has important implications for a very wide variety of human activities.

In the most immediate and practical sense, astronautics is a very large engineering job. Equipments and facilities, often requiring substantial advances over current practice, must be designed and built. Severe environmental conditions and demands for high reliability over very long periods of essentially unattended operation will require uncompromising thoroughness and extensive testing. Bold imagination and painstaking attention to detail must be the twin hallmarks of engineering for space flight.

Engineering action can be founded only on scientific knowledge. The scientist must support the engineer with adequate data on the many aspects of space environments, and with a growing body of fundamental knowledge. Astronautics can furnish unparalleled new opportunities to the scientist to explore and understand man and his universe. Space vehicles can carry the scientist's instruments—and eventually the scientist himself—to regions otherwise not accessible to gather information otherwise unattainable. Astronautics presents the life sciences with two particularly challenging prospects: the problem of maintaining human existence outside the narrow living zone at the Earth's surface, and the possibility of encountering living things on other planets.

The departure of man and his machines from the very Earth itself is bound to have a profound influence on human thought and the general view of man's place in the scheme of things. His findings on other worlds can be expected to influence the broad development of philosophy to a degree comparable to that resulting from the invention of the telescope, whereby man discovered that he was not actually the center of the universe. Perhaps astronautics will show man that he is also not alone in the universe.

These and other aspects of the revolutionary nature of extraterrestrial exploration have prompted serious theological discussion. Implications of space flight with respect to Christian principles are matters of lively interest. As early as September 1956, Pope Pius XII formally stated that space activities are in no way contradictory to Church doctrine.[10]

The statesman, endeavoring to promote world peace, can see both a hope and a threat in astronautics. International cooperation in space enterprises could help to promote trust and understanding. Astronautics can provide physical means to aid international inspection and, thereby, can help in the progress toward disarmament and the prevention of surprise attack. Astronautics can also lead to military systems which, once developed and deployed, may make hopes of disarmament, arms control, or inspection immeasurably more difficult to fulfill.

International cooperation in astronautics is imperative simply as a matter of efficiency. Scientific space exploration cannot reasonably be done in isolated national packages. The long history of astronomy as an international science clearly demonstrates the point. Observation of natural celestial bodies, which (as viewed from the Earth) are permanent and relatively slow moving, has required the closest kind of international collaboration. The observation (not to mention creation and retrieval) of artificial celestial bodies, transient and fast moving, will place even heavier, more urgent, demands on international cooperation.

There is also obvious need for international cooperation in such matters as agreement on radio frequency allocations for space vehicles; and on rights of access to, and egress from, national territories for recovery of vehicles, particularly in cases of accidentally misplaced landings of manned vehicles.

Astronautics raises substantial questions of law, both international and local. The important issues of international agreement on space access and use must be afforded the most thoughtful sort of attention. Legal factors of a more conventional nature are also inherent in astronautics. Large tracts of real estate will be required for operations and testing, for example. The physical needs of astronautics are, therefore, a matter of important concern also to the civic planner—somewhat in the manner of airports and marine facilities.

Astronautics is inherently a high-cost activity that will clearly have an important impact on Government expenditures, taxes, corporate profits, and personal incomes. For the future, it may hold considerable promise of substantial economic benefits—astronautics is an entire new industry.

In astronautics lies the possibility of improved performance in important public and commercial service activities: weather forecasting, aids to navigation and communication, aerial mapping, geological surveys, forest-fire warning, iceberg patrol, and other such functions.

For national security and military operations, astronautics holds more than new means for implementing standard operations like reconnaissance and bombing. It suggests novel capabilities of such magnitude that entirely new concepts of military action will have to be developed to exploit them. As an obvious parallel, airplane technology has come a long way since Kitty Hawk; but the military thinking that determines the role of aircraft in national arsenals has also come a very long way indeed. A similar companion development of technology and military concepts can be expected to occur in astronautics also—but we can no longer afford the comparatively leisurely pace of adjustment that characterized the thinking about aircraft.

Astronautics has another important military dimension if "military" is interpreted in the broad sense of an organized, trained, and disciplined activity. It is hard to conceive of space exploration efforts such as manned voyages to Mars, involving many months of hazard and hardship, being undertaken by any but a "military" type of organization.

Astronautics is the sort of activity in which anyone can find means for satisfying personal participation. The work of amateurs in optical and radio observation of satellites has already been of great value, and there is no reason to believe that amateur activities in astronautics will not take a place alongside and within such vigorous hobbies as amateur radio and amateur astronomy.

Astronautics is bound to have an important impact on education. The broad nature of the problems to be faced will require not only specialists, but minds trained to cut across and exploit various classical disciplines.

C. Current State of Space Technology

The physical assets of the United States and other countries in astronautics now reside in large measure in military activities. More specifically, these current assets lie mainly in ballistic-missile programs.

The space flight capabilities that can be built on ballistic-missile assets are very extensive, indeed. The greatest of these are derivable from ICBM hardware. Adaptation of these vehicles to accommodate specially developed additional equipment will permit us to do the following:

(*a*) Orbit satellite payloads of 10,000 pounds at 300 miles altitude.
(*b*) Orbit satellite payloads of 2,500 pounds at 22,000 miles altitude.
(*c*) Impact 3,000 pounds on the Moon.
(*d*) Land, intact, 1,000 pounds of instruments on the Moon.
(*e*) Land, intact, more than 1,000 pounds of instruments on Venus or Mars.
(*f*) Probe the atmosphere of Jupiter with 1,000 pounds of instruments.
(*g*) Place a man, or men, in a satellite orbit around the Earth for recovery after a few days of flight.

All of these and other feats can be accomplished by starting with basic rocket vehicles now in development in this country. None, however, will come from the ballistic-missile programs directly. All require additional work of a very substantial nature. With diligence and reasonable luck, the overall rocket machinery necessary to attempt any of these flights could be available in a few years—probably less than five. Rocket capabilities of roughly the same order can reasonably be assigned to the U. S. S. R. in this period also—perhaps more.

At some point in the next 5 years the effect of larger engine developments should make itself felt; so we can already look forward to the day when the payloads listed above will be five to ten times greater.

These basic vehicle capabilities reflect the status of (*a*) chemical propulsion systems (mainly systems using liquid oxygen and kerosene); (*b*) vehicle structural materials, design techniques, and fabrication methods; and (*c*) vehicle flight stabilization (autopilot) methods. Much remains to be done in these fields, but a solid footing has been established.

Other fields, in various states of advancement, must also be considered, however, to obtain a full view of our current standing.

While knowledge of the space environment is uncertain in

many respects, no important barriers stand in the way of unmanned flights. So far as manned flight is concerned, no such definite statement can be made—partly because the requirements for human survival are much more severe than for instruments, partly because instrument flights can be one-way while manned flights must be round-trips, and partly because the human risk attached to mistakes is great for manned flight.

Accurate guidance of space vehicles over interplanetary ranges may require improvement in current knowledge of basic astronomical constants.

Performance of current guidance and navigation equipment is probably adequate for most satellite and lunar flight missions, but probably not at all adequate for flights to the planets.

Systems to control the orientation of space vehicles during free flight are (except for spin stabilization) in the untested and uncertain category.

Communication between space vehicles and Earth stations is rather easy to maintain in satellite or lunar flights. Communication as far as Mars seems reasonably attainable, but at much greater distances current possibilities become questionable.

Observation and tracking of vehicles will also be comparatively easy on satellite and lunar flights, and possible, with decreasing precision, to about the same ranges that apply for low-capacity communication systems—Venus and Mars, or beyond—with today's technology.

Techniques for high-speed penetration of the atmosphere, based on ballistic-missile re-entry developments, are adequate for nondestructive landings of instruments on Earth, Mars, or Venus. Methods are available for withstanding the shock of landing on planets or on the surface of the Moon.

Internal power sources now available are probably adequate for low-power applications over extended periods of time, or for short-time operation at moderate power levels. Sources suitable for large amounts of power over long periods of time are yet to be developed. Supply of internal power is one of the major problem areas.

A problem area that cannot be overemphasized is that of reliability. It is utterly meaningless to talk about flights to Mars if the equipment to be sent there has no reasonable probability of continuing to work for the duration of the flight and for a useful period after arrival. Keeping modest amounts of equipment working, unattended, for many months is possible, but it requires good knowledge of the environment, careful design, and extensive testing.

Electrical propulsion systems that can provide continuous thrust in the space environment are being investigated. All of these require extremely large amounts of electrical power —thousands of horsepower—making the already important problem of internal power supply even more prominent by consideration of such devices. Until one of these electrical propulsion systems is developed, all flights in space will be unpowered ballistic flights, with perhaps occasional spurts of corrective thrust from conventional chemical rockets.

The broad needs for sustained manned flight in space can be stated rather simply: Large launching rockets, extensive and highly reliable space-vehicle equipments, a great deal of experimentation and study, and, above all, actual flight trials in manned satellites or similar vehicles.

D. Action Considerations

A brief appreciation of the above remarks on technological status would be just this: Large-scale space exploration is imminently feasible with the beginnings now in hand; its actual accomplishment will require a great deal of work.

The value of a vigorous space program—and the urgent need for one—rests on many considerations, including moral, economic, scientific, defense, and international cooperation. The various objectives pertinent to these areas are closely related in themselves and in the kind of devices needed to implement them. A class of device that is "scientific" when it is originally planned may be "military" when it becomes a reality, and vice versa. There are, of course, always distinctions that can be drawn on the basis of primary intent. This is a matter of human choice—the machinery itself may not be changed much by shifts in purpose.

In trying to assess the value of a given space effort it is important to recall that new devices and methods must be examined not only for what they can do better than existing devices, but also for what they can do that now cannot be done at all. It is quite likely that astronautical devices will be found to be very poor competitors for many current conventional devices; much of the benefit is likely to lie in the doing of new kinds of things through astronautics. Practical realization of these ends is dependent upon our recognizing the unique capabilities in the field and developing (inventing, perhaps) the applications that make them useful. This is true in all the possible areas of interest. A space weapon will make no real contribution to national defense unless it is accompanied by a clear concept of useful employment—an innovation in military science must be sought in some cases. The physical capability for worldwide live television by satellite relays is a hollow thing without a complex set of plans and agreements that put suitable receiving sets in the homes on the ground and attractive program material into the transmitter. The opportunity to put scientific instruments into space is of only minor importance if it is not adequately supported by attention to basic theory and laboratory research on the ground.

Some of the unique opportunities that seem to lie in astronautics and are of obvious importance include the ability to do the following:

Carry scientific instruments out of the atmosphere and away from the Earth's magnetic field to permit greatly improved observations of remote regions of space, to advance basic understanding of some of the great fundamental questions about the nature of the universe and its large-scale processes.

Carry scientific instruments, and eventually people, to the planets and other bodies in the solar system for direct exploration of their physical make-up and, perhaps, of their indigenous life forms.

Permit studies of the behavior and evolution of terrestrial biological specimens in environments grossly different from that on Earth.

Enhance understanding of the physical properties of the Earth by viewing it from the vantage point of space to supplement surface observations.

Provide meteorological observations of global scope to improve understanding of weather processes, with a long-range hope not merely of better forecasting but eventually of some form of weather control.

Provide useful navigation beacons with global coverage independent of weather and time of day.

Provide large-scale radio broadcasting facilities, long-range point-to-point communications without elaborate, slowly constructed ground facilities; reliable communication at low power levels with fixed or vehicular stations; and point-to-point communication over long ranges at low power levels to conserve the already crowded radio frequency spectrum.

Conduct detailed military reconnaissance of vast areas in a short period of time.

Conduct international inspection of vast areas in a short period of time.

Make possible various forms of interference with military attack-warning systems.

Deliver nuclear weapons from remote regions of space.

Opportunities for international inspection through astronautics could be of incalculable value to world peace and security. The military potentialities are vast—and they can be exploited for good or ill by any power with the resources and will to try. The more scientific-seeming applications may, in the long run, exert a more profound influence on the future than any of the others. The massive importance of early research in nuclear physics in the laboratory and observatory is now apparent beyond doubt.

Notes

[1] Ley, W., Rockets, Missiles, and Space Travel, The Viking Press, Inc., New York, 1957.

[2] Dornberger, W. R., V-2, The Viking Press, New York, 1954.

[3] Randolph, Maj. J. R., What Can We Expect of Rockets? Army Ordnance, vol. XIX, No. 112, January-February 1939.

[4] See footnote 2.

[5] Dornberger, W. R., The Lessons of Peenemünde, Astronautics, vol. 3, No. 3, March 1958, p. 18.

[6] See footnote 5.

[7] Goddard, R. H., A Method of Reaching Extreme Altitudes, Smithsonian Institution publication No. 2540, 1919.

[8] Krieger, F. J., A Casebook on Soviet Astronautics, The RAND Corp., Research Memorandum RM-1760, June 21, 1956; A Casebook on Soviet Astronautics—Part II, RM-1922, June 21, 1957.

[9] Krieger, F. J., Behind the Sputniks: A Survey of Soviet Space Science, Public Affairs Press, Washington, D. C., 1958.

[10] Address of Pope Pius XII to the International Astronautical Congress, Castel Gandolfo, September 20, 1956. See L'Osservatore Romano, September 22, 1956.

PART 2

TECHNOLOGY

2

Space Environment

A. The Solar System

The salient known physical data on the principal objects of interest in the solar system are given in table 1. Many other, more detailed, characteristics of the planets such as the eccentricities and inclinations of their orbits, the inclination of their axes, their densities, albedoes, etc., are available in standard books on astronomy.

For a general appreciation of the environment of space travel the following characteristics of the solar system should be noted:

All of the nine planets move around the Sun in the same direction on near-circular orbits (ellipses of low eccentricity).

The orbits of the planets all lie in nearly the same plane (the *ecliptic*). The maximum departure is registered by Pluto, whose orbit is inclined 17° from the ecliptic.

One astronomical unit (a. u.), the mean distance of the Earth from the Sun, is 92,900,000 miles in length. The diameter of the solar system, across the orbit of its remotest member (Pluto), is about 79 a. u., or 7,300 million miles.

The four inner planets—Mercury, Venus, Earth, and Mars—are relatively small, dense bodies. These are known as the "terrestrial" planets.

The next four in distance from the Sun—Jupiter, Saturn, Uranus, and Neptune—sometimes called the major planets or the giant planets, are all relatively large bodies composed principally of gases with solid ice and rock cores at unknown depths below the visible upper surfaces of their atmospheres. Little is known about Pluto.

TABLE 1.—*Physical data on principal bodies of solar system*

Body	Mean distance from Sun (Earth's distance equals 1.00, or 1 a. u.)	Mass (Earth's mass equals 1.00)	Diameter (st. mi.)	Gravitational force at solid surface (g.'s)	Intensity of sunlight at mean distance (intensity at Earth = 1.0)	Length of day	Length of year	Number of moons
Sun	—	329,000	864,000	[1]	—			
Mercury	0.39	~.05	3,100	~0.3	6.7	88 days	88 days	0
Venus	.72	.82	7,500	.91	1.9	(?)	225 days	0
Earth	1	1	7,920	1.00	1	24 hours	365 days	1
Mars	1.52	.11	4,150	.38	.43	24.6 hours	1.9 years	2
Jupiter	5.2	317	87,000	[2]	.037	10 hours	12 years	12
Saturn	9.5	95	71,500	[2]	.011	do	29 years	9
Uranus	19.2	15	32,000	[2]	.0027	11 hours	84 years	5
Neptune	30	17	31,000	[2]	.0011	16 hours	165 years	2
Pluto	39	.8	(?)	(?)	.0006	(?)	248 years	0 (?)
Moon	1.00	.012	2,160	.17	1	27 days		0

[1] No solid surface.

[2] Location of solid surface (below the thousands of miles depth of dense atmospheric gases covering these planets) is not known; hence, surface gravity figures are meaningless for the 4 giant planets.

The general disposition of planetary orbits is illustrated in figure 1.

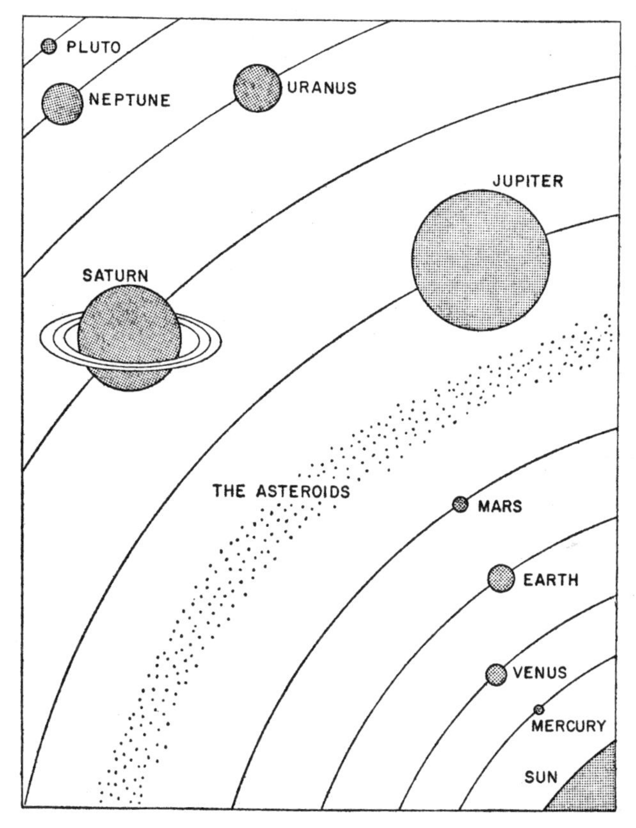

FIG. 1. The Solar System

B. The Sun

In astronomical terms, the Sun is a "main sequence" star of spectral type G-zero with a surface temperature of about 11,000° F.[1] Although a medium-small star, it is over a thousand times as massive as Jupiter, and over 300,000 times as massive as the Earth. Its energy output, as light and heat, is extremely constant, probably varying no more than about 0.5 percent from the average value.[2] However, it is much more variable in its production of ultraviolet radiation, radio waves (solar static), and charged particles. At infrequent intervals, extremely intense solar outbursts of charged particles (cosmic rays) have been observed. The most recent of these outbursts, which occurred on February 23, 1956, and lasted about 18 hours, resulted in a peak intensity of ionizing radiation above the atmosphere estimated at about 1,000 times the normal value.[3]

All usable forms of energy on the Earth's surface, with the exception of atomic and thermonuclear energy, are directly or indirectly due to the storing or conversion of energy received from the Sun.

C. The Planets

MERCURY

The planet closest to the Sun, Mercury, is difficult to observe because of its proximity to that body; hence, our knowledge of its physical characteristics is less accurate than for some of the other members of the solar system. Mercury has no moon, and its mass is not known with precision, but is about one-twentieth that of the Earth. This much is known, however: It is a small rocky sphere, about half again as large as the Earth's Moon, and it always keeps the same side turned toward the Sun. The sunlit half is thus extremely hot, probably having maximum surface temperatures as high as 750° F., while the side in perpetual darkness is extremely cold, cold enough to retain frozen gases, with temperatures approaching absolute zero. Mercury is not known to have any atmosphere, nor would a permanent gaseous envelope be expected to occur under the conditions existing on the planet. Its rocky surface is probably similar to that of our Moon.

VENUS

Even less is known with confidence about the surface conditions on Venus. Therefore, many statements about it are necessarily more speculative than definitive. In dimensions and mass it is slightly smaller than the Earth, but no astronomer has ever seen its solid surface, since its dense and turbulent atmosphere, containing white particles in suspension, is opaque to light of all wavelengths. Neither free oxygen nor water vapor has been detected on Venus, but carbon dioxide is abundant in its atmosphere, as determined by spectrographic analysis of the light reflected from the upper reaches of its visible cloud deck. On the basis of all the available evidence, it may be presumed that the surface of Venus is probably hot, dry, dusty, windy, and dark beneath a continuous dust storm; that the atmospheric pressure is probably several times the normal barometric pressure at the surface of the Earth; and that carbon dioxide is probably the major atmospheric gas, with nitrogen and argon also present as minor constituents.[4]

MARS

Much more complete information is available about Mars, but many questions about surface conditions still remain unanswered. With a diameter halfway between that of the Moon and the Earth, and a rate of revolution and inclination of Equator to orbital plane closely similar to those of Earth, it has an appreciable atmosphere and its surface markings exhibit seasonal changes in coloration. Its white polar caps, appearing in winter and vanishing in summer, are apparently thin layers of frozen water (frost) fractions of an inch to several inches in thickness. The atmospheric pressure at the surface has been estimated at 8 to 12 percent of Earth sea-level normal, and the atmosphere is believed to consist largely of nitrogen. No free oxygen has been detected in its atmosphere.[5] Nothing definite is known about the presence or absence of marked differences in the altitude of the terrain. The "climate" would be similar to that of a high desert on Earth to an exaggerated degree (about 11 miles high, in fact) with noon summer temperatures in the Tropics reaching a maximum of perhaps 80° to 90° F., but falling rapidly during the evening to reach a minimum before dawn of about −100° F. The interval between two successive close approaches of Earth and Mars is slightly over 2 years. At opposition, that is, when the two planets lie in the same directon from the Sun, the approximate distance between Earth and Mars ranges from 35 million to 60 million miles.

Bleak and desertlike as Mars appears to be, with no free oxygen and little, if any, water, there is rather good evidence of life.

The seasonal color changes, from green in spring to brown in autumn, suggest vegetation. Recently Sinton has found spectroscopic evidence that organic molecules may be responsible for the Martian dark areas.[6] The objections raised concerning differences between the color and infrared reflectivities of terrestrial organic matter and those of the dark areas on Mars have been successfully met by the excellent work of Prof. G. A. Tikhov and his colleagues of the new Soviet Institute of Astrobiology.[7] Tikhov has shown that arctic plants differ in infrared reflection from temperate and tropical plants, and an extrapolation to Martian conditions leads to the conclusion that the dark areas are really Martian vegetable life.

Although human life could not survive without extensive local environmental modifications, the possibility of a self-sustaining colony is not ruled out.

THE GIANT PLANETS

The four members of this group of planets (Jupiter, Saturn, Uranus, Neptune) have so many characteristics in common that they may well be treated together. They are all massive bodies of low density and large diameter. They all rotate rapidly. Because of their low densities (0.7 to 1.6 times the density of water) and on the basis of spectral information, they all are thought to have a "rock-in-a-snowball" structure —that is, a small dense rocky core surrounded by a thick shell of ice and covered by thousands of miles of compressed hydrogen and helium. Methane and ammonia are also known to be present as minor constituents.[8] Because of the low intensities of solar radiation at the distances of the giant planets, temperatures at the visible upper atmospheric surfaces range from −200° to −300° F. A number of the satellites of Jupiter, Saturn, and Neptune are larger than the Earth's Moon, and some may be as large as Mercury. Although reliable physical data on these satellites are lacking, it is possible that they might be somewhat more hospitable for space flight missions than the planets about which they orbit.

PLUTO

Almost nothing is known about this most distant member of the known solar system except its orbital characteristics and the fact that it is extremely cold, with a small radius and a mass about 80 percent that of the Earth.

D. Moon

The Moon is about 240,000 miles from the Earth, and its diameter is about 2,160 miles, a bit more than one-fourth the diameter of the Earth. The mass of the Earth is about 81.5 times that of the Moon.

The Moon has no appreciable atmosphere, and its surface is probably dry, dust-covered rock. On the basis of terrestrial experience it would be expected that this rocky surface is far from uniform in chemical composition and physical arrangement.

The face of the Moon is covered with many large craters, the origin of which is still a matter of debate. Mountains on the moon are higher than those on Earth, presumably because they are free from weathering. A Soviet astronomer recently reported observation of an erupting volcano on the Moon.[9] Whether or not the observations actually support the stated interpretation has been questioned by some authorities.

E. Asteroids

In addition to the planets and their moons, there is a group of substantial bodies known as *asteroids* in the solar system, more or less concentrated in the region between the orbits of Mars and Jupiter. It is possible that these chunks of material may be the shattered remains of one or more planets.

Most of the asteroids have dimensions of some miles, but quite a few are as much as 100 miles across. The largest, Ceres, is nearly 500 miles in diameter, and irregular in shape.

Some asteroids come within a few million miles of the Earth from time to time.

F. Comets

Comets are very loose collections of material that sweep into the inner regions of the solar system from space far beyond the orbit of Pluto. Some return periodically; some never do. Their bodies consist of rarefied gases and dust, and their heads are thought to be frozen gases or "ices."

G. Meteorites

The Earth receives a large quantity of material from surrounding space in the form of meteoritic particles. Most of these are decomposed in the upper atmosphere, but some reach the Earth's surface.

These particles enter the Earth's atmosphere with velocities of 7 to 50 miles per second, producing visible light streaks called *meteors*. Estimates of numbers, sizes, and speeds of incoming meteorites are based in part on optical observation of meteors, and in part on radio-wave reflections from the ionization trails left by meteorites. Data about smaller particles are deduced from other effects, such as sky glow at twilight.

Estimates, based on various assumptions, of the total volume of incoming meteoritic material range from 25 to 1 million tons per day. A very recent estimate made by the Harvard College Observatory favors a value of 2,000 tons per day.[10] Soviet scientists announced in August 1958 that their satellite data indicated an estimate of 800,000 to 1 million tons per day.[11]

The vast range of variability in these estimates is due in part to uncertainties in densities of meteoritic material, inadequate experimental techniques, too few observations, and incomplete theoretical bases for interpreting observations.

Information about meteoritic input to the Earth's atmosphere is very uncertain. The meteoritic content of other space regions is largely an open question, pending direct experimentation with space vehicles.

H. Micrometeorites and Dust

The smallest dust particles, called *micrometeorites,* are concentrated for the most part in the *ecliptic,* the plane of the Earth's orbit, and since they originate as cometary refuse they may also be found distributed along the orbits of comets.[12]

Evidence that cosmic dust is concentrated in the plane of the ecliptic comes from observations of a faint tapered band of light that can be seen at twilight extending up from the horizon centered along the ecliptic. This band of light, which can be photoelectrically traced through the complete night sky, is called the *zodiacal light.*

The layer of small meteoritic particles must extend from the Sun well beyond the orbit of the Earth, being concentrated toward the ecliptic or fundamental plane of the solar system. Further, this dust cloud must be continuously being resupplied by cometary wastage and possibly by material from asteroid collisions. It is at the same time being drained off by the action of solar radiation which causes the particles to spiral in toward the Sun. It has been estimated that in 60 million years all particles smaller than 1 millimeter in diameter, starting nearer to the Sun than the orbit of Mars, would reach the Sun due to this effect.

I. Radiation and Fields

At the surface of the Earth, man and his machines are sheltered by the atmosphere from many radiations that exist in space. These radiations include X-rays, steady ultraviolet radiation, and cosmic rays.

At times of solar flares there may be great quantities of radiation less energetic than cosmic rays, in addition to the particles now known to be trapped in the Earth's geomagnetic field.

Cosmic radiation is a general term for high-speed particles from space. About 80 percent are protons, which are nuclei of hydrogen atoms carrying a single positive charge, with the remainder consisting of a number of other subatomic particles.

Corpuscular cosmic rays, although "hard" or highly energetic radiation, are not so numerous as other corpuscular radiations within the solar system. The IGY Explorer satellites observed an encircling belt of high-energy radiation extending upward from a height of a few hundred miles. This belt ranges from at least 65° north to about 65° south latitude, although the radiation is most intense in the equatorial region.

Explorer IV data indicate that radiation intensity increases several thousand times between 180 and 975 miles, with a rapid rise beginning at about 240 miles. The level of radiation may reach as much as 10 roentgens per hour—enough to deliver an average lethal dose in 2 days to an unshielded human being.

Results from Sputnik III, which did not travel quite so far out in space but went to higher northern and southern latitudes, seem roughly compatible with the Explorer results. Tentative data from Pioneer indicate a rapid decay of radiation intensity with increasing distance beyond about 17,000 miles from the Earth.

Since the Earth's magnetic field, which stores and traps the particles comprising the radiation belt, decreases in strength with increasing distance from the Earth, it ultimately becomes too weak to store a significant amount of radiation.

Trapping of electrons or protons in the Earth's magnetic field is not expected to be of importance in very high latitudes, and, therefore, the radiation hazard will be much less to vehicles launched near the poles than to those traversing the lower-latitude radiation belt.

The apparent dose rate at altitudes between about 300 and 400 miles lies within the accepted AEC steady-state tolerance level for human beings of 300 milliroentgens per week (1.79 milliroentgens per hour). Therefore, this radiation belt does not interdict low-altitude manned satellites. It does imply that manned satellites orbiting at altitudes greater than 300 to 400 miles would require some shielding, the weight of shielding increasing up to the greatest altitude for which we have fairly firm information (roughly 1,200 miles). Beyond this altitude, the radiation levels are uncertain, but it is expected that at some altitude a maximum must be reached after which the dosage rates should diminish.

Depending upon the extent of the equatorial radiation belt, manned space flights could use the technique of leaving or returning to the Earth via the polar regions, or could penetrate directly through the radiation belt with adequate shielding to protect human beings during the transit.

The Sun, in addition to its steady radiations, delivers great outbursts with radiation levels as much as 1,000 times normal at the times of large solar flares, which appear about once every 3 years. Smaller increases occur almost daily.

At present, it is not possible to predict exactly when these large solar flares will occur. A man in a vehicle above the Earth's protective atmosphere, if exposed to the radiation from such a flare during its entire period of activity, might absorb enough radiation to make him ill (not immediately, but within a week or two). Since our knowledge is incomplete as to the peak intensities of solar flares, their frequency of occurrence, and their duration, the risk to human beings from this source cannot now be assessed with any real accuracy.

The constant radio noise background emitted by the Sun is also greatly enhanced during solar flares. This solar noise is reinforced somewhat by radio noise from distant galaxies. Some localized areas of far space radiate very intensely. Such a source is in the region of the Crab nebula.

J. Beyond the Solar System

Although not a theater of operations in the first phases of the space age, the larger setting in which the solar system itself figures is worthy of mention.

The nearest neighbor of our solar system is the star system Alpha Centauri, a bright object in the southern sky at a distance of about 4 light-years. (Pluto is 5½ light-hours from the Sun.) Alpha Centauri is a double star whose two main components orbit about one another. A third star called "Proxima" is also associated with this system and is actually at the present time the star closest to our solar system. ("Proxima" itself may also be doubled.)

It is not known whether the Alpha Centauri system has any planets, but observations of some other nearby stars, e.g., 61 Cygni, indicate, from wobbles in their motion, the possible presence of orbiting dark bodies with masses comparable to Jupiter's. There is, then, what might be considered indirect evidence for the existence of other planetary systems. Within 20 light-years of the Sun there are known to be about 100 stars with possibly two or three planetary systems, if the interpretation of the "wobbles" is correct. Kuiper estimates on the basis of the ratio of the masses of components of double stars that not more than 12 percent of all stars may have planetary systems.[13] When we realize that there are some 200 billion stars in our galaxy, this would give 1 to 10 billion with planetary systems. It seems reasonable to speculate that out of this vast number there surely must be some systems with earth-like planets, and that on some of these planets life similar to our own may have evolved.[14-17]

With our present state of knowledge, however, communication with such planetary systems is a matter of speculation only. When we recall that our galaxy is some 100,000 light-years in diameter, the Sun being an insignificant star some 30,000 light-years from the galactic center, circling in an orbit of its own every 200 million years as the galaxy rotates, we realize that even trying to visualize the tremendous scale of the universe beyond the solar system is difficult, let alone trying to attempt physical exploration and communication. Nor is the interstellar space of the galaxy the end, for beyond are the millions of other galaxies all apparently rushing from one another at fantastic speeds; and the limits of the telescopically observable universe extend at least 2 billion light-years from us in all directions.

Notes

[1] Hoyle, F., Frontiers of Astronomy, Harper & Bros., New York, 1955.

[2] Roberts, W. O., The Physics of the Sun, the Second International Symposium on the Physics and Medicine of the Atmosphere and Space, San Antonio, Tex., November 12, 1958.

[3] Schaefer, H. J., Appraisal of Cosmic-Ray Hazards in Extra Atmospheric Flight, Vistas in Astronautics, Pergamon Press, 1958.

[4] Dole, S. H., The Atmosphere of Venus, The RAND Corp., Paper P-978, October 12, 1956.

[5] Kuiper, G. P. (editor), The Atmospheres of the Earth and Planets, The University of Chicago Press, Chicago, Ill., 1952.

[6] Sinton, W. M., Further Evidence of Vegetation on Mars, presented at the meeting of the American Astronomical Society, Gainesville, Fla., December 27-30, 1958.

[7] Tikhov, G. A., Is Life Possible on Other Planets? Journal of

the British Astronomical Association, vol. 65, No. 3, April 1955, p. 193.
[8] Urey, H. C., The Planets, Their Origin and Development, Yale University Press, New Haven, Conn., 1952.
[9] The Moon Is Not Dead, USSR, No. 3 (30), 1959, p. 40.
[10] Whipple, F. L., The Meteoritic Risk to Space Vehicles, Vistas in Astronautics, Pergamon Press, p. 115.
[11] Nazarova, I. N., Rocket and Satellite Investigations of Meteors, presented at the fifth meeting of the Comité Spéciale de l'Année Géophysique Internationale, Moscow, August 1958. Nazarova has very recently revised the estimate of meteor influx downward by a factor of about 1,000.
[12] Beard, D. B., Interplanetary Dust Distribution and Erosion Effects, American Astronautical Society, Preprint No. 58-23, August 18-19, 1958.
[13] Kuiper, G. P., The Formation of the Planets, Journal of the Royal Astronautical Society of Canada, vol. 50, No. 2, p. 167.
[14] Jeans, Sir James, Life on Other Worlds, A Treasury of Science, edited by H. Shapley, Harper & Bros., New York, 1954.
[15] Shapley, H., Of Stars and Men: The Human Response to an Expanding Universe, Beacon Press, Boston, 1958.
[16] Planets: Do Other "Humans" Live? Newsweek, vol. LII, No. 20, November 17, 1958, p. 23.
[17] Haldane, J. B. S., Genesis of Life, The Earth and Its Atmosphere, edited by D. R. Bates, Basic Books, Inc., New York, 1957.

3

Trajectories and Orbits

A. Fundamental Types of Trajectories and Orbits

The terms *trajectory* and *orbit* both refer to the path of a body in space. *Trajectory* is commonly used in connection with projectiles and is often associated with paths of limited extent, i.e., paths having clearly identified initial and end points. *Orbit* is commonly used in connection with natural bodies (planets, moons, etc.) and is often associated with paths that are more or less indefinitely extended or of a repetitive character, such as the *orbit* of the Moon around the Earth. In discussions of space flight, both terms are used, with the choice usually dependent upon the nature of the flight path. Thus we speak of *trajectories* from the Earth to Moon, and of satellite *orbits* around the Earth.

The basic types of paths in space are determined by the gravitational-attraction properties of concentrated masses of material and the laws of motion discovered by Newton.

Virtually all major members of the solar system are approximately spherical in shape; and a spherical body will produce a force of attraction precisely like that of a single mass point located at the center of the body. Therefore, the fundamental problem is that of motion under the gravitational influence of a mass concentrated at a point.

Two general and several special types of paths are possible under the gravitational influence of a point mass. The two main types are illustrated in figure 1.

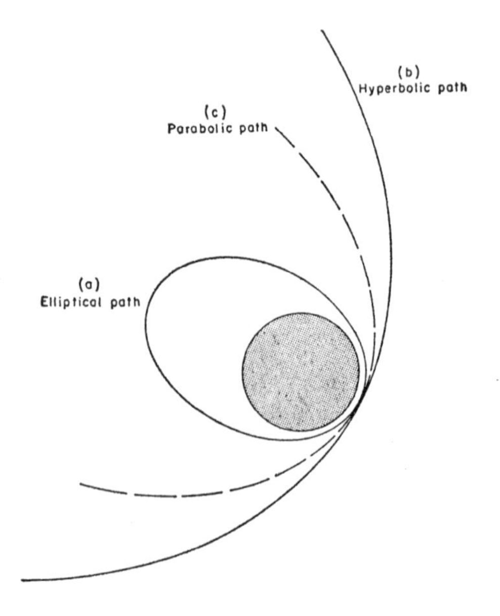

FIG. 1. Types of paths

Figure 1a is an *elliptical orbit*—the familiar artificial Earth satellite orbit. Figure 1b is a *hyperbolic orbit*—the kind that will characterize the start of an interplanetary flight. The elliptical orbit is closed on itself and would be traversed repetitively. The hyperbolic orbit is open, extending to infinity. Separating these two cases is a special one—the *parabolic orbit* —similar in general appearance to the hyperbolic. The parabolic orbit is the borderline case between open and closed orbits and therefore identifies the borderline condition between space vehicles that are tied to paths (elliptical) in the general vicinity of their parent planet and those that can take up paths (hyperbolic) extending to regions remote from their parent planet. For any of these orbits the vehicle's velocity will be greatest at the point of nearest approach to the parent body, and it will be progressively less at more remote points.

B. Escape Velocity

The type of path that will be taken up by an unpowered space vehicle starting at a given location will depend upon its velocity. It will take up an open-ended path if its velocity equals or exceeds *escape velocity;* escape velocity is, by definition, that velocity required at a given location to establish a parabolic orbit. Velocities greater than escape velocity result in hyperbolic orbits. Lower velocities result in closed elliptical orbits —the vehicle is tied to the neighborhood of the planet.

Since it essentially separates "local" from "long distance" flights, escape velocity is clearly a primary astronautical parameter. The exact value of this velocity is dependent upon two factors: (*a*) The mass of the parent planet, and (*b*) the distance from the center of the planet to the space vehicle. Escape velocity increases as the square root of the planet's mass, and decreases as the square root of the distance from the planet's center. The speeds required for escape directly from the surfaces of various bodies of interest are listed in table 1. These escape velocity requirements are a measure of the difficulty of departure from these bodies.

The projection speed required to escape directly from the Earth's surface is about 36,700 feet per second. If a vehicle takes up unpowered flight (end of rocket propulsion) at an altitude of, say, 300 miles, it requires the somewhat lesser speed of 35,400 feet per second to escape into interplanetary space. This reduction in required velocity has, of course, been obtained at the expense of the energy expended in lifting the vehicle to an altitude of 300 miles.

TABLE 1.—*Surface escape velocity*

	Feet per second		*Feet per second*
Mercury	13,600	Mars	16,700
Venus	33,600	Asteroid Eros	~50
Earth	36,700	Jupiter	197,000
Moon	7,800		

C. Satellite Orbits

The elliptical orbits generated by velocities below escape velocity are the type followed by artificial satellites, as well as by all the planets and moons of the solar system.

The period of the satellite—the time required to make one full circuit—is dependent upon the mass of the parent body and the distance across the orbit at its greatest width (the length of the *major axis*). The period is less if the parent body is more massive—the Earth's Moon moves more slowly than similarly placed moons of Jupiter. The period gets longer as the length of the major axis increases—the period of the Moon, with a major axis of about 500,000 miles, is much longer than those of the first artificial satellites, with major axes of about 9,000 miles.

The velocity required to establish a satellite at an altitude of a few hundred miles above the Earth is about 25,000 feet per second. This required *orbital velocity* is less at greater altitudes. At the distance of the Moon it is only about 3,300 feet per second.

D. Lunar Flight

The gravitational attraction of the Moon affords some assistance to a vehicle on an Earth-Moon flight. However, the Moon is so far removed that this assistance is only enough to reduce the required launching velocity slightly below escape velocity.

E. Interplanetary Flight

To execute a flight to one of the other planets, a vehicle must first escape from the Earth. Achieving escape velocity, however, is only part of the problem; other factors must be considered, particularly the Sun's gravitational field and the motion of the Earth about the Sun.

Before launching, the vehicle is at the Earth's distance from the Sun, moving with the Earth's speed around the Sun—about 100,000 feet per second. Launching at greater than Earth escape velocity results in the vehicle's taking up an independent orbit around the Sun at a velocity somewhat different from that of the Earth. If it is fired in the same direction as the Earth's orbital motion, it will have an independent velocity around the Sun greater than that of the Earth. It will then take up an orbit such as A, figure 2, which moves farther from the Sun than the Earth's orbit; the vehicle could, if properly launched, reach the outer planets Mars, Jupiter, and so forth. The minimum launch velocities required to reach these planets are given in table 2.

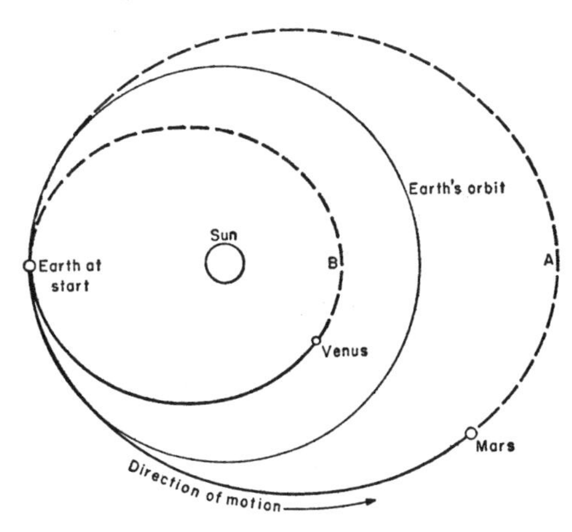

FIG. 2. Interplanetary trajectories

If the vehicle is launched "backward," or against the Earth's velocity, it will assume an independent velocity less than that of the Earth and move on an orbit like B, figure 2, so that it could reach the inner planets Venus and Mercury.[1]

To reach the more distant portions of the solar system requires that the vehicle take up a velocity, relative to the Sun, that is considerably *greater* than that of the Earth. A large launch velocity is required to produce this excess (after a good deal of it has been absorbed by the Earth's gravitational field). On the other hand, to travel in close to the Sun requires that the vehicle take up a velocity, relative to the Sun, that is considerably *less* than that of the Earth. A large launch velocity is this time required to cancel out the component of vehicle velocity due to the Earth's motion, and again much of the launch velocity is absorbed by the Earth's gravitational field. Thus, as seen from table 2, it is almost as hard to propel a vehicle in to Mercury as it is to propel it out to Jupiter.

TABLE 2.—*Minimum launch velocities, with transit times, to reach all planets*

Planet	Minimum launching velocity (feet per second)	Transit time
Mercury	44,000	110 days
Venus	38,000	150 days
Mars	38,000	260 days
Jupiter	46,000	2.7 years
Saturn	49,000	6 years
Uranus	51,000	16 years
Neptune	52,000	31 years
Pluto	53,000	46 years

The velocities in table 2 are minimum requirements, and lead to the transit times shown. Higher velocities will reduce transit times.

F. Escape from the Solar System

If a vehicle reaches escape velocity with respect to the Sun it will leave the solar system entirely and take up a trajectory in interstellar space. Starting from the surface of the Earth, a launch velocity of about 54,000 feet per second will lead to escape from the solar system. The course of the vehicle will be a parabola, with the Sun at its focus, until eons later it is deflected by some star or other body.

The flight capabilities that become available as the total velocity potential of a ballistic rocket vehicle increases are illustrated in summary form in figure 3.

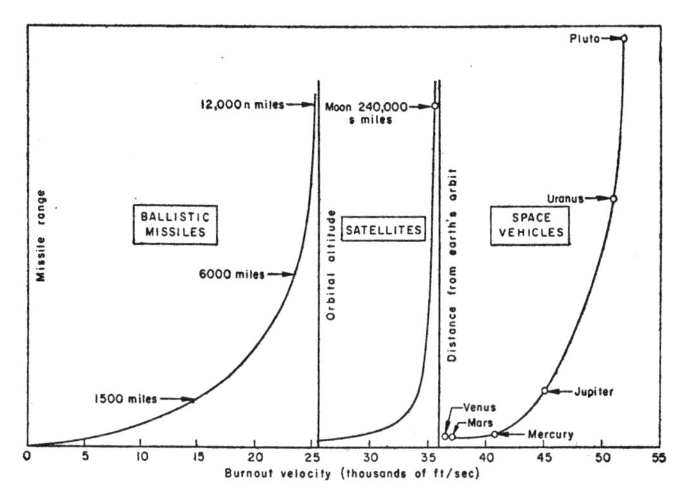

FIG. 3. Velocity requirements for ballistic missile and space flight

G. Powered Trajectories in Space

Once a vehicle is in space, moving at high velocity, say in a satellite orbit, it requires no further propulsion to stay aloft. Its flight path can, however, be very appreciably influenced and great increases in velocity imparted by very small forces acting over long periods of time. The fact that useful results can be derived from small thrusts in space—thrusts that would be entirely insignificant on the Earth—leads to interest in unique propulsion systems based on electrical accelerator principles. One kind of application of particular interest involves the use of heavy conventional propulsion systems to develop orbital velocity (say, 25,000 feet per second) and then to build up the remaining 12,000 feet per second to reach escape velocity by a low-thrust electrical system. [2, 3]

H. Velocities Near That of Light

As the velocity of a space vehicle nears that of light (not likely to be achieved in the foreseeable future), the effects of relativity theory enter into the situation. Of particular interest is the so-called "time dilatation" effect predicted by this theory—and supported by experimental evidence in the physics of high-speed particles.

Briefly, the predicted effect is as follows: Consider two men, A and B, of identical age, say, 20 years old. A will remain at home on the Earth, and B will undertake a voyage in space at a speed very near that of light and eventually return to Earth. The total duration of the voyage will be different, as measured by the two men, the exact amount of the difference depending upon how close B's vehicle approached the speed of light.

As an example, suppose B took a round trip to the vicinity of a nearby star at a speed very near that of light (about 186,000 miles per second). It would appear to A that the trip took, say, 45 years—he would be 65 years old when his friend returned. To B, however, the trip might appear to take about 10 years, including a year or so for acceleration to flight speed and deceleration for the return landing—he would be 30 when he returned.

Different values of vehicle speed will lead to widely different time disparities. By approaching ever closer to the speed of light, B could take more extended trips that would last millions of years in earthtime, but still appear to him to take only a few years.

Achievement of near-light velocities would require stupendous amounts of propulsion energy—nothing less than complete conversion of matter into usable energy will do.[4-6]

In addition to the fact that no presently foreseeable propulsion scheme will deliver the required quantities of energy, there are also problems of a very severe and uncertain nature concerning environment. The traveler at speeds approaching that of light will find himself immersed in a grossly altered natural environment and will also face the problems of carrying with him a source of extremely intense radiation—in whatever form his propulsion system may take.

Notes

[1] The time of day of launching is also different for flights to regions closer or farther from the Sun than the Earth. For areas farther away, seeking to use the Earth's orbital and rotational speeds to best advantage, the launching must be from the dark side of the Earth. On the other hand, a point that is in daylight will be moving with a speed opposite to the Earth's orbital motion and such a point would be chosen for missions nearer the Sun. Hence, to Mars by night and to Venus by day.

[2] Boden, R. H., The Ion Rocket Engine, North American Aviation, Rocketdyne Division, Rept. No. R-645, August 26, 1957.

[3] Stuhlinger, E., Space Travel with Electric Propulsion Systems, Army Ballistic Missile Agency, November 11, 1958.

[4] Gamow, G., One, Two, Three—Infinity, Mentor MS 97, Paperbound Books, c. 1947.

[5] Von Handel, P. F., and H. Knothe, Relativistic Treatment of Rocket Kinematics and Propulsion, Air Force Missile Development Center, Rept. No. AFMDC TR 58-3, January 1958.

[6] Haskins, J. R., Notes on Asymmetrical Aging in Space Travel, Quarterly Research Review No. 18, Army Rocket and Guided Missile Agency, Rept. No. 2A18, September 1, 1958.

4

Rocket Vehicles

A. General Description of Rocket Vehicles

The principal elements of any rocket-powered flight vehicle are the *rocket engine,* to provide the propulsive force; the *propellants* consumed in the rocket engine; the *airframe,* to contain the propellants and to carry the structural loads; and the *payload,* including any special equipment such as guidance or communication devices.

The rocket engine provides the propulsive forces to accelerate the vehicle by ejecting hot gaseous material at very high speeds through a nozzle. The initial source of the ejected material is the propellant carried in the vehicle in either liquid or solid form. The propellant may be converted to hot gas for ejection by one of a number of possible heating processes in the engine, such as chemical combustion, nuclear fission, etc.

Vertical takeoff from the Earth requires a thrust force that exceeds the weight of the complete missile by some 30 to 50 percent (a thrust-to-weight ratio of 1.3 to 1.5). For easiest engine operation the thrust produced during the entire propulsive period is usually constant, causing the vehicle to be accelerated at a progressively higher rate as the vehicle weight diminishes due to propellant consumption.

In rocket vehicles intended to reach velocities of interest in astronautics, the largest fraction of the missile weight is devoted to the propellants, and the largest volume to the storage of these propellants.

The propellant tanks, and the supporting structure which carries the structural loads imposed during flight and ground handling, comprise the *airframe* of the flight vehicle. The material in the airframe is considered "dead weight," since it does not contribute directly to the production of thrust or to the useful payload. Rather, the dead weight imposes a limitation on the maximum velocity that a given rocket can achieve —even with no payload.

Another factor contributing significantly to the total dead weight of a vehicle and restricting its maximum performance is any unused propellant trapped in the propulsion system (rocket engine, plumbing, and tanks) at thrust cutoff. In liquid-propellant rockets, two propellant fluids are stored in separate tanks which should be emptied at very nearly the same instant.[1] The engine will stop when either propellant is exhausted, and the remaining portion of the other propellant will be trapped as residual dead weight.

The flight velocities required for astronautics far exceed those obtainable with a single rocket unit using conventional propulsion techniques, regardless of the size of the rocket. The *multistage* rocket can provide adequate velocities, however. On this type of vehicle, one rocket (or more) is carried to high speed by another rocket, to be launched independently when the first rocket is exhausted. If, for example, the first stage reaches a terminal velocity of 10,000 feet per second and launches a second stage also capable of developing 10,000 feet per second, the net terminal velocity of the second stage will be 20,000 feet per second.

Staging can be extended to include three, four, and more stages to develop higher velocities. The total velocity attained is the sum of the individual contributions of each stage. A practical difficulty will generally restrict the number of stages that can be profitably employed, since the weight of structure required to connect the stages tends to increase dead weight and defeat the purpose of staging.

Control of the flight path of a rocket-propelled vehicle is achieved by altering the direction of engine thrust by one of several methods, including swiveling the engine itself. Velocity control is provided through termination of all rocket thrust at the exact time the desired velocity is reached.

B. Vehicle Parameters

The performance of a rocket is determined largely by the rocket-propellant combination and the total amount of usable propellants.[2] The performance of propellants is characterized by the *specific impulse,* a measure of the number of pounds of thrust produced per pound of propellant consumed per second. The unit of specific impulse is lb/lb/sec, or, more simply, seconds. The velocity that a vehicle can attain is directly proportional to the specific impulse of its propellants, all other things being equal. For example, if a given rocket reaches a velocity of 10,000 feet per second with propellants giving a specific impulse of 250 seconds (a typical current value), an increase of 10 percent in specific impulse (to 275 seconds) would increase the attainable velocity to 11,000 feet per second.

The dependence of performance upon the *propellant fraction*—the fraction of the total vehicle weight accounted for by usable propellants—and the interdependence of specific impulse and propellant fraction are more complex relationships. The maximum velocity increases rapidly as the propellant fraction is made larger, as shown in figure 1. For a given velocity to be achieved, the required propellant fraction is greatly affected by the performance of the propellant combination, as indicated by the three values of specific impulse given in this figure. (The relationship between the propellant fraction and the *mass ratio,* the ratio of takeoff weight to weight at propellant exhaustion, is given by the lower scale of figure 1.) The very high propellant fractions associated with

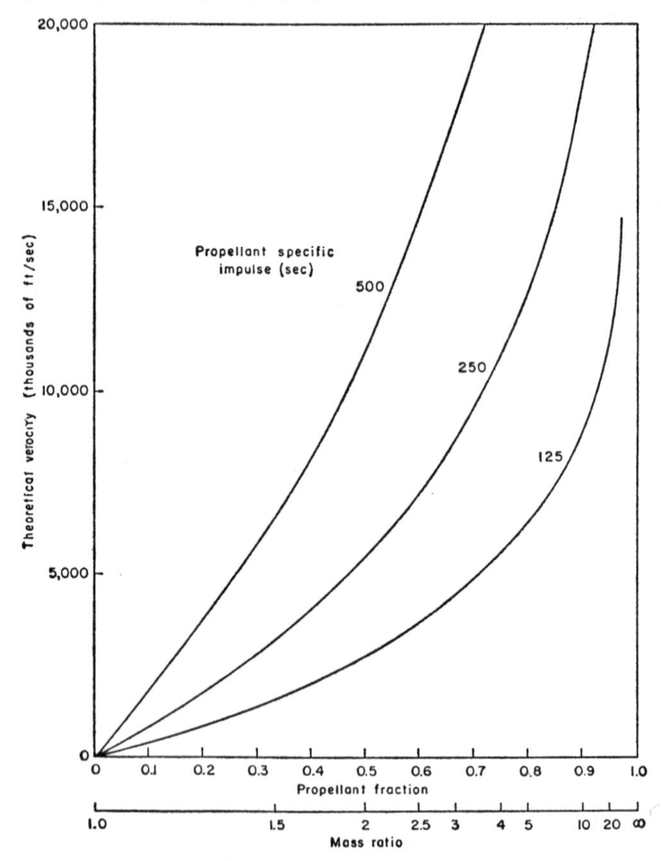

FIG. 1. Velocity characteristics of rocket vehicles
(*Single stage*)

high velocities can only be achieved by severely reducing to a minimum all components that contribute to the weight at propellant exhaustion, including the payload.

The gross weight at takeoff of a rocket vehicle to propel a given payload to a given flight velocity is determined, then, by propellant performance, the minimum practical fraction of dead weight, the number of stages, and the size of the payload to be carried.

Improved materials, design techniques, and component miniaturization have led to large reductions in the fraction of dead weight associated with large rockets. The progressively improving state of the art is graphically illustrated in figure 2, from a value of 0.25 for the German V-2 development in 1939,[3] to 0.21-0.16 for the United States Viking rockets in 1948-51,[4] to a reported value near 0.08-0.06 for current large rocket developments in the United States.[5]

The gross weight (and thrust) required for a specified flight objective is very nearly proportional to the size of the payload to be carried. To double the payload, the gross weight —and the thrust—must also be about doubled for similar de-

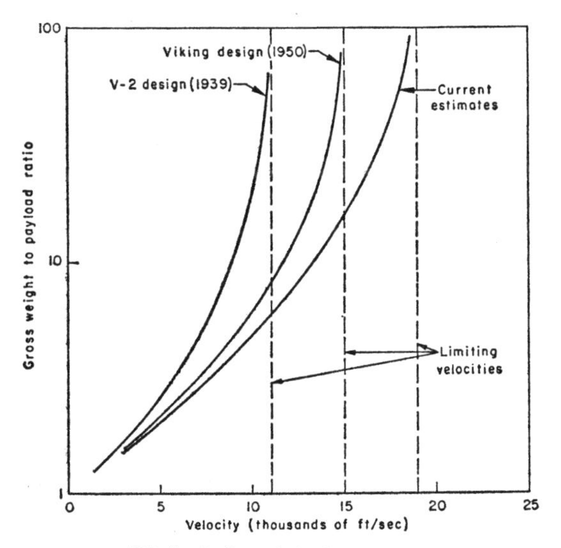

FIG. 2. Rocket vehicle characteristics (*Single stage*)

signs and vehicle configurations. This is the predominant reason for development of large-thrust engines. Additional stages and improved propellants may be used to obtain increased performance from existing rocket-booster components.

Notes

[1] Greenwood, T. L., A High Accuracy Liquid Level Measuring System, Army Ballistic Missile Agency, Redstone Arsenal, Ala., Rept. DTI-TR-1-58, July 29, 1958.

[2] A more complete discussion of propellants appears in section 6.

[3] The Missile A-4 Series B as of January 2, 1945, translation, General Electric DF-71369.

[4] Viking Design Summary, RTV-N-12a, Glenn L. Martin Co., ER-6534, August 1955.

[5] Astronautics and Space Exploration, hearings before the Select Committee on Astronautics and Space Exploration, 85th Cong., 2d sess., on H. R. 11881, April 15 through May 12, 1958.

5

Propulsion Systems

A. Principles of Operation of Rocket Engines

The only known way to meet space-flight velocity requirements is through the use of the rocket in one of its several forms.

Rocket *thrust* is the reaction force produced by expelling particles at high velocity from a nozzle opening. These expelled particles may be solid, liquid, gaseous, or even bundles of radiant energy. The engine's ability to produce thrust will endure only so long as the supply of particles, or *working fluid,* holds out. Expulsion of *material* is the essence of thrust production; and without material to expel, no thrust can be produced, regardless of how much energy is available.

Because of this fundamental fact, a prime criterion for rating rocket performance is specific impulse, which provides an index of the efficiency with which a rocket uses its supply of propellant or working fluid for thrust production. For gaseous working fluids, specific impulse can be increased by (1) attaining higher temperatures in the combustion chamber and (2) increasing the proportion of lighter gases, preferably hydrogen, in the exhaust.

The other important factor in assessing the merit of a propulsion system in a given application is the weight of engine and working fluid container required, since these weights influence achievable *propellant fraction.*

B. Types of Rocket Engines

Rocket engines are distinguished by the type of mechanism used to produce exhaust material. The simplest "engine" is a compressed air bottle attached to a nozzle—the exhaust gas is stored in the same form as it appears in the exhaust. Ejection of compressed air, or other gas, from a nozzle is a perfectly satisfactory rocket operation for some purposes.

The most common rocket engine is the chemical type in which hot exhaust gases are produced by chemical combustion. The chemicals, or *propellants,* are of two types, *fuel* and *oxidizer,* corresponding to gasoline and oxygen in an automobile engine. Both are required for combustion. They may be solid or liquid chemicals.

In other types of rockets no chemical change takes place within the engine, but the working fluid may be converted to a hot gas for ejection by the addition of heat from a nuclear reactor or some other energy source.

These and other variations of the rocket engine are discussed below.[1]

C. Solid-Propellant Rocket

In the solid-chemical rocket, the fuel and oxidizer are intimately mixed together and cast into a solid mass, called a *grain,* in the combustion chamber (figure 1).[2] The propellant grain is firmly cemented to the inside of the metal or plastic case, and is usually cast with a hole down the center. This hole, called the *perforation,* may be shaped in various ways, as star, gear, or other, more unusual, outlines. The perforation shape and dimension affects the burning rate or number of pounds of gas generated per second and, thereby, the thrust of the engine.

After being ignited, usually by an electrical impulse, the propellant grain burns on the entire inside surface of the perforation. The hot combustion gases pass down the grain and are ejected through the nozzle to produce thrust.

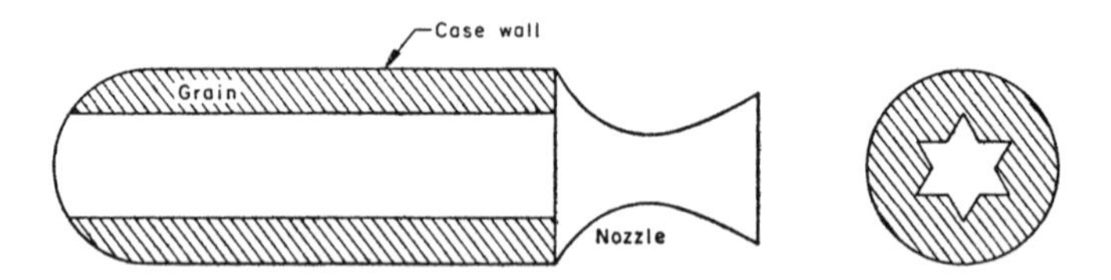

FIG. 1. Schematic of solid-propellant rocket

The propellant grain usually consists of one of two types of chemical. One type is the *double-base,* which consists largely of *nitroglycerine* and *nitrocellulose.* It resembles smokeless gunpowder. The second type, which is now predominant, is the *composite* propellant, consisting of an oxidizing agent, such as *ammonium nitrate* or *ammonium perchlorate* intimately mixed with an organic or metallic fuel. Many of the fuels used are plastics, such as *polyurethane.*

A solid propellant must not only produce a desirable specific impulse, but it must also exhibit satisfactory mechanical properties to withstand ground handling and the flight environment. Should the propellant grain develop a crack, for example, ignition would cause combustion to take place in the crack, with explosion as a possible result.

It can be seen from figure 1 that the case walls are protected from the hot gas by the propellant itself. Therefore, it is possible to use heat-treated alloys or plastics for case construction. The production of light-weight, high-strength cases is a major development problem in the solid-rocket field.

Since nozzles of solid rockets are exposed to the hot gas flowing through them, they must be of heavy construction to retain adequate strength at high temperature. Special inserts are often used in the region of the nozzle throat to protect the metal from the erosive effects of the flowing gas.

For vehicle guidance it is necessary to terminate thrust sharply upon command. This may be accomplished with solid rockets by blowing off the nozzle or opening vents in the chamber walls. Either of these techniques causes the pressure in the chamber to drop and, if properly done, will extinguish the flame.

The specific impulse of various solid-propellant rockets now falls in the range of 175 to 250 seconds. The higher figure of 250 applies to ammonium perchlorate-based propellants.[3]

D. Liquid Bipropellant Chemical Rockets

The common liquid rocket is bipropellant; it uses two separate propellants, a liquid fuel and liquid oxidizer. These are contained in separate tanks and are mixed only upon injection into the combustion chamber. They may be fed to the combustion chamber by pumps or by pressure in the tanks (figure 2).

Propellant flow rates must be extremely large for high-thrust engines, often hundreds of gallons per second. Pump-fed systems may require engines delivering several thousand horsepower to drive the pumps.[4] This power is usually developed by a hot-gas turbine, supplied from a gas generator which is actually a small combustion chamber. The main rocket propellants can be used for the gas generator, although, as in the case of the V-2 and the Redstone, a special fuel like hydrogen peroxide can be used for this purpose.[5]

The pressure-feed system eliminates the need for pumps and turbines; however, the high pressure required in the tanks, perhaps 500 pounds per square inch, leads to heavier structures, thus adding dead weight to the vehicle that may more than offset the weight saved by removing the pumping system.[6] On the other hand, removal of pumping equipment may increase reliability.

The walls of the combustion chamber and nozzle must be protected from the extremely high gas temperature. The method most commonly used is to provide passage in the nozzle wall through which one of the propellants can be circulated. In this way the walls are cooled by the propellant, which is later burned. This technique is referred to as regenerative cooling.[7]

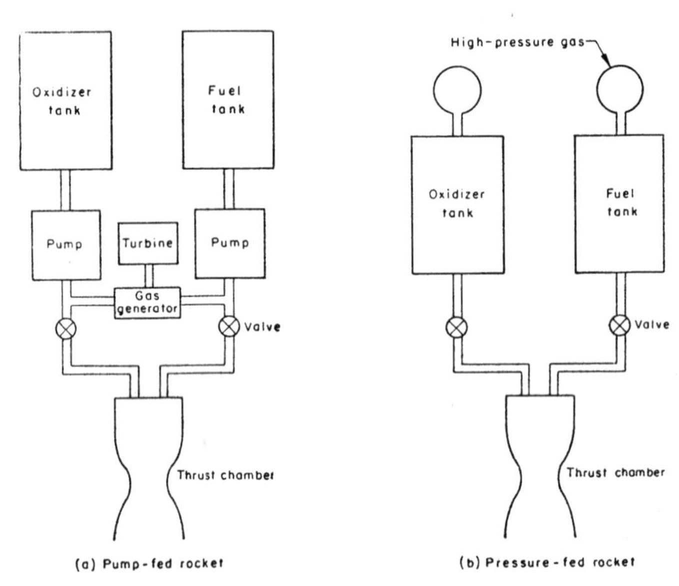

FIG. 2. Schematic of liquid-propellant rocket

Thrust termination is easily accomplished with the liquid rocket by simply shutting the propellant valves; however, this operation must be precisely timed and controlled. The amount of thrust delivered can be controlled by adjusting the rate of propellant flow.

E. Liquid Monopropellant Rocket

Certain liquid chemicals can be made to form hot gas to produce thrust by *decomposition* in a rocket chamber. The most common such *monopropellant* is hydrogen peroxide. When this liquid is passed through a platinum *catalyst* mesh it decomposes into hot steam and oxygen. These gases can then be ejected to develop thrust.

Engines of this kind have comparatively low specific impulse, but have the advantage of simplicity, require only one tank in the vehicle, and can be readily turned on and off. Since they are adaptable to repetitive operation they find application in various control systems where efficiency of propellant use is of minor importance.[8]

F. Nuclear Rockets

Research and development on the use of a nuclear reactor as a rocket energy source is currently being carried out in Project Rover.[9] The nuclear rocket does not use any combustion process. Rather, the hot exhaust gas is developed by passing a working fluid through a fission reactor (figure 3). Liquid hydrogen is the propellant most often considered for a nuclear rocket because it yields the lightest exhaust gas possible. The hydrogen could be stored in liquid form in a single tank and forced into a reactor by a pump. After being heated in the reactor, it would be exhausted through a conventional rocket nozzle to obtain thrust.[10]

Other methods of using the fission reactor have been proposed to avoid the severe materials problem attendant on transferring heat to the gas directly by the extremely hot reactor walls. One device would place gaseous fissile material in the center of an open reactor, retaining it in position by magnetic means. Then the propellant gas would be heated by radiation from the hot gaseous fissile material, without the interposition of a solid wall. The feasibility of such a device is still a subject of investigation.[11]

Specific impulse figures for conventional nuclear rockets may be as high as 1,200 seconds.[12-15]

It has also been proposed that atom (fission) bombs of limited power be exploded below a space vehicle to push it along. Heavy construction would be required to protect the interior of the vehicle from blast and radiation effects.[16]

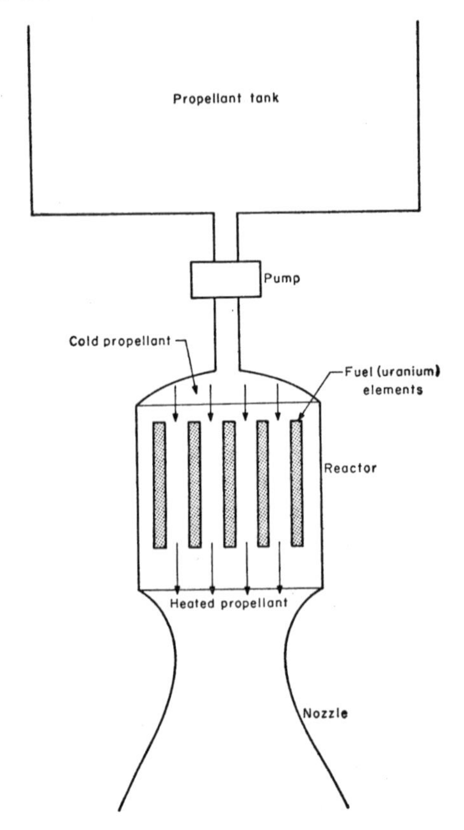

FIG. 3. Nuclear rocket

G. Thermonuclear Rockets

Harnessing thermonuclear reactions to obtain power is a subject of continuing interest throughout the world. The United States effort is being conducted under Project Sherwood.[17] It is reasonable to suppose that a thermonuclear reactor could be used as an energy source for a rocket in ways not basically different from those suggested for nuclear reactors.[18]

The use of thermonuclear reactors or other advanced schemes for propulsion (plasma rockets, ion rockets) involves phenomena of the type falling under the general term, *magnetohydrodynamics:* study of the behavior of ionized gases acted upon by electric and magnetic fields. Magnetohydrodynamics is one of the very active fields of research in engineering today.[19, 20]

H. Solar Propulsion

A number of schemes have been proposed to use radiation from the Sun to obtain propulsive power for a space ship. Although the energy density of solar radiation in space is rather small in comparison with the tremendous power of chemical launching rockets, it can be useful for propulsion in "open" spaces. Once a vehicle is well away from the Earth or other planetary body, or is established in a satellite orbit, a very small amount of thrust will serve to alter or accelerate its flight significantly.

Solar propulsion schemes fall into two categories. In one, the *radiation pressure* of solar rays would be used to supply thrust on a large, lightweight surface attached to the space ship—quanta (bundles) of radiant energy, or *photons,* are the working materials of such a rocket. Thus, propellant is supplied in an endless stream from the Sun and no storage tanks are required on the vehicle. This device has been called a *solar sail* (figure 4).[21] The other approach is to use the solar rays to heat hydrogen gas, which is then expelled through a nozzle to produce thrust.

In both of these approaches the weight of mechanism relative to the thrust obtainable is likely to be so large as to severely limit the usefulness of solar propulsion.

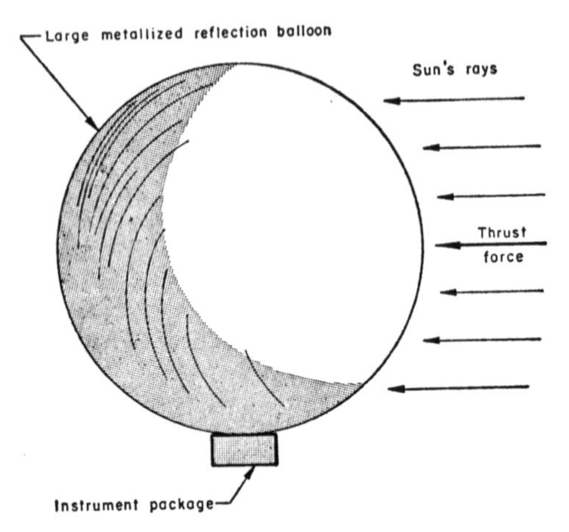

FIG. 4. Solar sail

I. Ion Propulsion

In the various devices for *ion propulsion,*[22] each molecule of propellant (usually assumed to be an alkali metal, notably cesium) is caused to have an electric charge; that is, the propellant is ionized (figure 5). This might be accomplished by passing the propellant over heated metal grids. It is then possible to accelerate the charged molecules, or ions, to very high velocities through a nozzle by means of an electric field. (Electrons are accelerated in a television picture tube in this fashion.) The performance of such an ion engine is very good, with values of specific impulse estimated to be as high as 20,000 seconds.[23] However, the amount of electric power required is very large, so weight of the power-generating equipment becomes a major obstacle to an efficient vehicle. It is contemplated that some type of nuclear fission (or fusion, farther in the future) could be used to supply the energy for the electric powerplant, although this step would still not eliminate the need for heavy electrical generators, unless direct conversion of fission to electrical energy in large quantities became practical.[24-26]

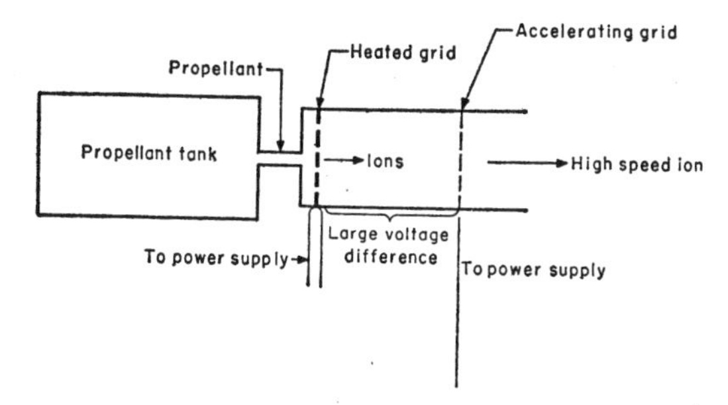

FIG. 5. Ion rocket engine

For example, an ion rocket offering 20,000 seconds of specific impulse, using cesium for the propellant, would require about 2,100 kilowatts of electric power to produce 1 pound of thrust, assuming good efficiency. Optimistic estimates of electric-power-supply weight indicate that the power unit would weigh about 8,500 pounds. The weight of the ion accelerator itself is small in comparison. Therefore, an ion rocket can accelerate itself only very slowly (about 1/10,000 of 1 g in this example).

J. Plasma Rocket

Another possible method for using electric power to operate a rocket engine uses electricity to heat the propellant directly by discharging a powerful arc through it. In this way, very high temperatures can be obtained, leading to high specific impulse, perhaps several thousand seconds, while avoiding the materials problem involved in heating a gas by passing it over a hot surface, as would be done, for example, in the conventional nuclear rocket. Such a device has been termed a *plasma rocket*.[27, 28] It, too, requires large quantities of electric power, about 150 kilowatts for each pound of thrust. While the sketch (figure 6) shows a conventional rocket nozzle, it may be possible to use magnetic fields to direct the jet, as the heated propellant has been ionized (made into a *plasma*) by the arc (another application of magnetohydrodynamics).

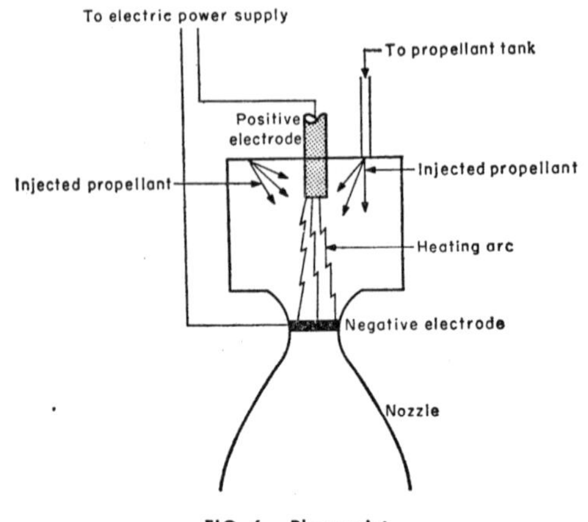

FIG. 6. Plasma jet

K. Photon Rocket

Several research workers have proposed that photons, that is, light or other radiation, might be generated and emitted from the rocket in a focused beam. A certain amount of momentum is associated with the photon beam, and thrust would be generated by such an engine. Such a system, however, would use energy very inefficiently, unless matter could be converted completely into energy.[29] For example, a large military searchlight is a photon rocket in a sense, but yields less than one ten-thousandth of a pound of thrust for a power consumption of 100 kilowatts.

L. Summary

There are two general measures of the performance of a rocket engine. One is the specific impulse, which will determine the amount of propellant that must be used to accomplish a given task. The second is the fixed weight of the engine, including the necessary tankage, power supply, and structure.

The chemical rocket engine is a fairly lightweight device. However, the specific impulse is not high. Solid and liquid propellants in present use deliver an impulse of about 250 seconds. The best liquid propellants so far conceived and evaluated yield an impulse of about 350 seconds. Certain solid propellants, proposed on the basis of theory alone, might yield 300 seconds. The fundamental theory of chemical binding energies precludes the possibility of any substantial gains over these numbers. Even some as-yet-undiscovered superfuel is unlikely to raise the specific impulse beyond 400 seconds or thereabouts.

The heat-transfer nuclear rocket is not limited by propellant binding energies, but by the temperature limitations of wall materials. Using hydrogen as a propellant, values of specific impulse of perhaps 1,200 seconds are feasible. Should gaseous containment of the fissioning fuel be possible, specific impulses of several thousand seconds might be achieved. This type of rocket engine appears very promising, and research on nuclear rockets and controlled thermonuclear power reactors may yield information useful to the construction of such a device.

The primary consideration in obtaining useful thrust from ion or plasma rockets is the construction of lightweight electric power supplies. A gross reduction in electrical generation equipment, as compared with the most advanced of present equipments, is required to make the electric rocket really interesting for flight in the solar system.

In any event, the electric rocket is likely to remain a low-thrust device. Therefore, large chemical or nuclear rockets would still be required to boost a space ship from the surface of the Earth.

No prospects are now apparent for realization of propulsion schemes of the "antigravity" variety, because the negation or reversal of the gravitational attraction of matter would violate basic physical laws as presently understood. Pending discovery of a new class of physical phenomena, the notion of antigravity now stands in a state similar to that of the perpetual motion machine.

M. Air-Breathing and Recoverable Boosters

The use of air-breathing engines, principally turbojets or turbofans, as first-stage missile boosters is certainly feasible.

However, for boosting large missiles (current ICBM's or larger) an air-breathing booster would be a relatively complex and expensive device, either composed of many jet engines already being developed for aircraft applications, or requiring the special development of large jet engines having several times greater thrust than conventional jet engines. Consequently, from an economic point of view, air-breathing boosters are of interest only in these cases where the operation permits the repeated use of the boosters (perhaps 10 times or more); thus, recoverable boosters might be of interest for possible future large-scale satellite-launching operations, but not for ICBM systems requiring a quick-reaction salvo capability. It is possible that, when all the operational factors are considered, the potential savings from using recoverable air-breathing boosters will turn out to be relatively small, perhaps considerably less than 20 percent.

Since the possible economic advantage of air-breathing recoverable boosters arises from the recoverability and reuse feature of these boosters, rather than from the fact that they are air-breathing, they are also worthy of consideration. For example, the rocket motor, and perhaps also the empty propellant tank, of a liquid rocket booster could be returned to Earth intact by parachutes, gliding wings, lifting jet engines, or by some combination of these. A reasonable compromise among the desirable qualities of reliability of launch and recovery, ease of return of booster to launch site, and minimum booster cost might result in a liquid rocket booster, whose rocket motor is returned to the launch site by relatively small turbojet engines; these jet engines would also assist the rocket booster during ascent through the lower atmosphere.

Notes

[1] See also section 6.

[2] Shafer, J. I., Solid-Rocket Propulsion, Jet Propulsion Laboratory, California Institute of Technology, external publication No. 451, April 10, 1958.

[3] Some Considerations Pertaining to Space Navigation, Aerojet-General Corp., Special Rept. No. 1450, May 1958, p. 18.

[4] Alexander, W. R., Mechanical Design Problems of Components in High Performance Spacecraft, News in Engineering, The Ohio State University Press, July 1958, p. 20.

[5] Sutton, George P., Rocket Propulsion Elements, John Wiley & Sons, New York, 1949, ch. 7.

[6] Corporal Propulsion System, Jet Propulsion Laboratory, California Institute of Technology, external publication No. 417, September 30, 1957.

[7] Atlas Propulsion System Background, North American Aviation, Rocketdyne Division, press release No. RS-4, March 10, 1958.

[8] Viking Design Summary, RTV-N-12a, Glenn L. Martin Co., Rept. ER-6534, August 1955.

[9] Outer Space Propulsion by Nuclear Energy, hearings before subcommittees of the Joint Committee on Atomic Energy, Congress of the United States, 85th Cong., 2d sess., January 22, 23, and February 6, 1958.

[10] Bussard, R. W., and R. D. DeLauer, Nuclear Rocket Propulsion, McGraw-Hill, New York, 1958.

[11] Bussard, R. W., Some Boundary Conditions for the Use of Nuclear Energy in Rocket Propulsion, American Rocket Society Preprint No. 690-58, 1958.

[12] Sutton, G. P., A Preliminary Comparison of Potential Propulsion Systems for Space Flight, a speech before the Wichita Section, American Rocket Society, June 30, 1957.

[13] Outer Space Propulsion by Nuclear Energy, hearings before subcommittees of the Joint Committee on Atomic Energy, Congress of the United States, 85th Cong., 2d sess., January 22, 23, and February 6, 1958; Col. J. T. Armstrong, p. 186.

[14] Outer Space Propulsion by Nuclear Energy, hearings before subcommittees of the Joint Committee on Atomic Energy, Congress of the United States, 85th Cong., 2d sess., January 22, 23, and February 6, 1958; R. Schreiber, p. 28.

[15] National Aeronautics and Space Act, hearings before the Special Committee on Space and Astronautics, U. S. Senate, 85th Cong., 2d sess., on S. 3609, pt. 1; A. Silverstein, p. 39.

[16] Outer Space Propulsion by Nuclear Energy, hearings before subcommittees of the Joint Committee on Atomic Energy, Congress of the United States, 85th Cong., 2d sess., January 22, 23, and February 6, 1958; Dr. S. Ulam, p. 47.

[17] Physical Research Program, hearings before the Subcommittee on Research and Development of the Joint Committee on Atomic Energy, February 1, 1958, p. 395.

[18] Bussard, R. W., Concepts for Future Nuclear Rocket Propulsion, Jet Propulsion, vol. 28, No. 4, April 1958, p. 223.

[19] Yoler, Y. A., Some Magnetohydrodynamic Problems in Aeronautics and Astronautics, Boeing Airplane Co., Document No. D1-7000-26, September 1958.

[20] Gauger, J., V. Vali, and T. E. Turner, Laboratory Experiments in Hydromagnetic Propulsion, Lockheed Aircraft Corp., Missile System Division.

[21] Garwin, R. L., Solar Sailing, Jet Propulsion, March 1958, p. 188.

[22] Ion Rockets and Plasma Jets, Air Force Office of Scientific Research, April 17, 1958.

[23] Willinski, M. I., and E. C. Orr, Project Snooper, Jet Propulsion, November 1958, p. 723.

[24] See footnote 23.

[25] Willinski, M. I., and E. C. Orr, Project Snooper: A Program for Reconnaissance of the Solar System with Ion Propelled Vehicles, North American Aviation, Rocketdyne Division, June 19, 1956.

[26] Outer Space Propulsion by Nuclear Energy, hearings before subcommittees of the Joint Committee on Atomic Energy, Congress of the United States, 85th Cong., 2d sess., January 22, 23, and February 6, 1958; T. Merkle, p. 560.

[27] Reid, J. W., The Plasma Jet: Research at 25,000° F., Machine Design, February 6, 1958.

[28] Outer Space Propulsion by Nuclear Energy, hearings before subcommittees of the Joint Committee on Atomic Energy, Congress of the United States, 85th Cong., 2d sess., January 22, 23, and February 6, 1958; A. Silverstein, p. 80.

[29] Huth, J. H., Some Fundamental Considerations Relating to Advanced Rocket Propulsion Systems, The RAND Corp., Paper P-1479, November 21, 1958.

6

Propellants

A. General Features of Rocket Propellants

Chemical propellants in common use deliver specific impulse values ranging from about 175 up to about 300 seconds. The most energetic chemical propellants are theoretically capable of specific impulses up to about 400 seconds.

High values of specific impulse are obtained from high exhaust-gas temperature, and from exhaust gas having very low (molecular) weight. To be efficient, therefore, a propellant should have a large heat of combustion to yield high temperatures, and should produce combustion products containing simple, light molecules embodying such elements as hydrogen (the lightest), carbon, oxygen, fluorine, and the lighter metals (aluminum, beryllium, lithium).

Another important factor is the density of a propellant. A given weight of dense propellant can be carried in a smaller, lighter tank than the same weight of a low-density propellant. Liquid hydrogen, for example, is energetic and its combustion gases are light. However, it is a very bulky substance, requiring large tanks. The dead weight of these tanks partly offsets the high specific impulse of the hydrogen propellant.

Other criteria must also be considered in choosing propellants. Some chemicals yielding high specific impulse create problems in engine operation. Some are not adequate as coolants for the hot thrust-chamber walls. Others exhibit peculiarities in combustion that render their use difficult or impossible. Some are unstable to varying degrees, and cannot be safely stored or handled. Such features inhibit their use.

Unfortunately, almost any propellant that gives good performance is likely to be a very active chemical; hence, most propellants are corrosive, flammable, or toxic, and are often all three. One of the most tractable liquid propellants is gasoline. But while it is comparatively simple to use, gasoline is, of course, highly flammable and must be handled with care. Many propellants are highly toxic, more so even than most war gases; some are so corrosive that only a few special substances can be used to contain them; some may burn spontaneously upon contact with air, or any organic substance, or in certain cases some common metals.

A rocket propellant must also be available in quantity. In some cases, to obtain adequate amounts of a propellant, an entire new chemical plant must be built.

B. Solid Chemical Propellants

Two sorts of solid propellant are in use: double-base and composite. The first consists of nitrocellulose and nitroglycerine, plus additives in small quantity. There is no separate fuel and oxidizer. The molecules are unstable, and upon ignition break apart and rearrange themselves, liberating large quantities of heat. These propellants lend themselves well to smaller rocket motors. They are often processed and formed by extrusion methods, although casting has also been employed.

In the composite propellant, separate fuel and oxidized chemicals are used, intimately mixed in the solid grain. The oxidizer is usually ammonium nitrate, potassium chlorate, or ammonium chlorate, and often comprises as much as four-fifths or more of the whole propellant mix. The fuels used are hydrocarbons, such as asphaltic-type compounds, or plastics. Because the oxidizer has no significant structural strength, the fuel must not only perform well but must also supply the necessary form and rigidity to the grain. Much of the research in solid propellants is devoted to improving the physical as well as the chemical properties of the fuel.

Ordinarily, in processing solid propellants the fuel and oxidizer components are separately prepared for mixing, the oxidizer being a powder and the fuel a fluid of varying consistency. They are then blended under carefully controlled conditions and poured into the prepared rocket case as a viscous semisolid. They then set in curing chambers under controlled temperature and pressure.

Solid propellants offer the advantage of minimum maintenance and instant readiness. However, the more energetic solids may require carefully controlled storage conditions, and may cause handling problems in the very large sizes. Because the rocket must always be carried about fully loaded, it must be protected from mechanical shocks or abrupt temperature changes that may crack the grain.

C. Liquid Chemical Bipropellants

Most liquid chemical rockets use two separate propellants: a fuel and an oxidizer. Typical fuels include kerosene, alcohol, hydrazine and its derivatives, and liquid hydrogen. Many others have been tested and used. Oxidizers include nitric acid, nitrogen tetroxide, liquid oxygen, and liquid fluorine. Some of the best oxidizers are liquified gases, such as oxygen and fluorine, which exist as liquids only at very low temperatures; this adds greatly to the difficulty of their use in rockets. Most fuels, with the exception of hydrogen, are liquids at ordinary temperatures.

Certain propellant combinations are *hypergolic;* that is, they ignite spontaneously upon contact of the fuel and oxidizer. Others require an igniter, although they will continue to burn when injected into the flame of the combustion chamber.

In general, the liquid propellants in common use yield specific impulses superior to those of available solid propellants. On the other hand, more complex engine systems are needed to transfer the liquid propellants to the combustion chamber. A list of solid- and liquid-propellants and their performance is given in table 1.

TABLE 1.—*Specific impulse of some typical chemical propellants*[1]

	Isp range (sec)
Propellant combinations:	
Monopropellants (liquid):	
Low-energy monopropellants	160 to 190
Hydrazine	
Ethylene oxide	
Hydrogen peroxide	
High-energy monopropellants:	
Nitromethane	190 to 230
Bipropellants (liquid):	
Low-energy bipropellants	200 to 230
Perchloryl fluoride—Available fuel	
Analine—Acid	
JP-4—Acid	
Hydrogen peroxide—JP-4	
Medium-energy bipropellants	230 to 260
Hydrazine—Acid	
Ammonia—Nitrogen tetroxide	
High-energy bipropellants	250 to 270
Liquid oxygen—JP-4	
Liquid oxygen—Alcohol	
Hydrazine—Chlorine trifluoride	
Very-high-energy bipropellants	270 to 330
Liquid oxygen and fluorine—JP-4	
Liquid oxygen and ozone—JP-4	

[1] Some Considerations Pertaining to Space Navigation, Aerojet-General Corp., Special Rept. No. 1450, May 1958.

Liquid oxygen—Hydrazine	
Super-high-energy bipropellants	300 to 385
Fluorine—Hydrogen	
Fluorine—Ammonia	
Ozone—Hydrogen	
Fluorine—Diborane	
Oxidizer-binder combinations (solid):	
Potassium perchlorate:	
Thiokol or asphalt	170 to 210
Ammonium perchlorate:	
Thiokol	170 to 210
Rubber	170 to 210
Polyurethane	210 to 250
Nitropolymer	210 to 250
Ammonium nitrate:	
Polyester	170 to 210
Rubber	170 to 210
Nitropolymer	210 to 250
Double base	170 to 250
Boron metal components and oxidant	200 to 250
Lithium metal components and oxidant	200 to 250
Aluminum metal components and oxidant	200 to 250
Magnesium metal components and oxidant	200 to 250
Perfluoro-type propellants	250 and above

Liquid oxygen is the standard oxidizer used in the largest United States rocket engines. It is chemically stable and non-corrosive, but its extremely low temperature makes pumping, valving, and storage difficult. If placed in contact with organic materials, it may cause fire or an explosion.

Nitric acid and nitrogen tetroxide are common industrial chemicals. Although they are corrosive to some substances, materials are available to contain these fluids safely. Nitrogen tetroxide must be protected because it boils at fairly low temperatures.

Liquid fluorine is a very-low-temperature substance, like liquid oxygen, and is highly toxic and corrosive as well. Furthermore, its combustion products are extremely corrosive and dangerous. For these reasons, there are many difficulties associated with the use of fluorine in testing and operating rocket engines.

Most other liquid fuels, with the exception of hydrogen, are similar in performance and handling. They are usually quite tractable substances. Hydrogen, however, is liquid only at extremely low temperatures—lower even than liquid oxygen; hence, it is very difficult to handle and store. Also, if allowed to escape into the air, hydrogen can form a highly explosive mixture. It is also a very bulky substance, only one-fourteenth as dense as water. Nevertheless, it offers the best performance of any of the liquid fuels.

D. Liquid Chemical Monopropellants

Certain unstable liquid chemicals which, under proper conditions, will decompose and release energy, have been tried as rocket propellants. Their performance, however, is inferior to that of bipropellants or modern solid propellants, and they are of most interest in rather specialized applications, as in small control-rockets. Outstanding examples of this type of propellant are hydrogen peroxide and ethylene oxide.[1]

E. Combinations of Three or More Chemical Propellants

The combining of more than two chemicals as propellants in rockets has never received a great deal of attention, and is not now considered advantageous. Occasionally a separate propellant is used to operate the generator which supplies the gas to drive the turbopumps of liquid rockets. In the V-2, for example, hydrogen peroxide was decomposed to supply the hot gas for the turbopumps, although the main rocket propellants were alcohol and liquid oxygen.

F. Free-Radical Propellants

When certain molecules are torn apart, they will give off large amounts of energy upon recombining. It has been proposed that such unstable fragments, called free radicals, be used as rocket propellants. However, these fragments tend to recombine as soon as they are formed; hence, the problem in their use is to develop a method of stabilization. Atomic hydrogen is the most promising of these substances: it might yield a specific impulse of about 1,200 to 1,400 seconds.[2]

G. Working Fluids for Nonchemical Rockets

Nuclear rockets must use some chemical as a working fluid or propellant, although no energy is supplied to the rocket by any chemical reaction. All the heat comes from the reactor. Because the prime aim is to minimize the molecular weight of the exhaust gas, liquid hydrogen is the best substance found so far and is not likely to be surpassed.

Another substance mentioned for use as a propellant in the nuclear rocket is ammonia. While offering only about one-half the specific impulse of hydrogen for the same reactor temperature because of its greater molecular weight, it is liquid at reasonable temperatures and is easily handled. Its density is also much greater than hydrogen, being about the same as that of gasoline.

Suitable propellants for use in electric propulsion devices are the easily ionized metals. The one most generally considered is cesium; others are rubidium, potassium, sodium, and lithium.

H. Unconventional Propellant Packaging

Some unconventional approaches in the propellant field include the following:

> Use of a liquid oxidizer with a solid fuel, the oxidizer being pumped through the perforation of the solid grain for burning.
>
> Sealing liquid propellant in small capsules, so that a liquid load can be handled and tanked as a mass of dry "propellant pills." [3]

Notes

[1] North American Aviation, Inc., press release NL-45, October 15, 1958.

[2] Outer Space Propulsion Using Nuclear Energy, hearings before subcommittees of the Joint Committee on Atomic Energy,

Congress of the United States, 85th Cong., 2d sess., January 22, 23, and February 6, 1958; Lt. Col. P. Atkinson, p. 145.

[3] Encapsulated Liquid Fuel Study Initiated, Aviation Week, vol. 69, No. 19, November 10, 1958, p. 29.

7

Internal Power Sources

A. Power Needs

All space vehicles will require some source of electrical power for communication equipment, instrumentation, environment controls, and so forth. In addition, vehicles using electrical propulsion systems—such as ion rockets—will have very heavy power requirements.

Current satellites and space probes have relatively low electrical power requirements—only a few watts. Bolder and more sophisticated space missions will lead to larger power needs. For example, a live television broadcast from the Moon may require kilowatts of power.[1] Over the distance from Earth to Mars at close approach, even a low-capacity communication link might easily require hundreds of kilowatts of power.[2] The power needs of men in space vehicles are less clearly defined, but can probably be characterized as "large." [3] Electrical propulsion systems will consume power at the rate of millions of watts per pound of thrust.

Power supply requirements cannot be estimated on the basis of average power demands alone. A very important consideration is the peak demand. For example, a radio ranging device may have an average power of only 2 watts, but it may also require a 600-watt peak. Unfortunately most foreseeable systems are severely limited in their ability to supply high drain rates; consequently they must be designed with a continuous capacity nearly equal to the peak demand.

A third important consideration is the voltage required. Voltage demand may be low for motors or high for various electronic applications. Furthermore, alternating current may be required or may be interchangeable with direct current. Transformations of voltage or shifts from direct to alternating current may be effected, but with a weight penalty.[4]

B. Batteries

The power source most readily available is the battery, which converts chemical to electrical energy. Table 1 summarizes the ultimate and the currently available performances of a few selected battery systems. The theoretical performance figures assume that all of the cell material enters completely into the electrochemical reaction.[5] These theoretical limits are unobtainable in practice because of the need for separators, containers, connectors, etc. The hydogen-oxygen (H_2–O_2) system refers to a fuel cell. Hydrogen and oxygen, stored under pressure, take the place of standard electrodes in a battery reaction, and about 60 percent of the heat of combustion is available as electrical energy.[6] The figures listed under "Currently available performance" in table 1 refer to long discharge rates (in excess of 24 hours) and normal temperatures.

TABLE 1.—*Electrochemical systems*

Type of cell	Limiting theoretical performance (watt-hours per pound)	Currently available performance	
		Watt-hours per pound	Watt-hours per cubic inch
Lead-acid	75	15	1.0
Nickel-cadmium	92	15	1.0
Zinc-silver	176	40	3.0
Hydrogen-oxygen	1,700	300	4.5

Batteries do not last long, do not operate well at low ambient temperatures or under heavy loads, and so are best suited to be storage devices to supply peak loads and to supplement some other prime source of energy.

Other characteristics of chemical batteries are: (1) They are essentially low-voltage devices, a battery pack being limited to about 10 kilovolts by reliability considerations; (2) some forms of radiation may injure them; and (3) many batteries form gas during charge and have to be vented, which is in conflict with the need for hermetic sealing to eliminate loss of electrolyte in the vacuum environment of space.

C. Solar Power

Solar energy arrives in the neighborhood of the Earth at the rate of about 1.35 kilowatts per square meter. This energy can be converted directly into electricity through the use of solar cells (solar batteries), or collected [7] to heat a working fluid which can then be used to run some sort of engine to deliver electrical energy.

Solar cells are constructed of specially treated silicon wafers, and are very expensive to manufacture. They cost about $100 per watt of power capacity.

Table 2 gives the performance at 80° F. of some current silicon cells produced by the Hoffman Electronics Corp. These cells generally operate at about 9 percent efficiency on the overall solar spectrum. On a weight basis they deliver about 30 watts per pound (bare). Possible hazards to solar cells include collisions with micrometeorites (which might have an effect similar to sandblasting) and various forms of solar radiation. Experiments on high-speed-impact phenomena have not yet clarified the extent of damage to be expected.[8] It has been estimated that solar cells should survive for many years in a solar radiation environment.[9] Surface cooling may be necessary for solar cells.

Satellites using solar cells must have batteries to store energy for use during periods of darkness. Solar cells and storage batteries in combination can be expected to weigh about 700 pounds per kilowatt of capacity.

The efficiency of solar cells can be improved. Of all the

energy received, part is effectively used in producing electrical energy, part is reflected (about 50 percent), and part actually passes through the cell, particularly the longer wavelengths. One obvious improvement would be to reduce the reflectivity of the cell; another would be to make the cells thinner.[10] Another possibility might be to concentrate solar energy through a lightweight plastic lens. It would also be desirable to develop cells for high-temperature operation.[11] Temperature control problems for solar cells have been investigated in the U. S. S. R. for satellite applications.[12]

TABLE 2.—*Performance of solar cells*

[Solar power available at the top of the Earth's atmosphere: 135 milliwatts per square centimeter]

Overall dimensions (centimeters)	Active area (square centimeters)	Output in full sunlight, with a matched load			
		Voltage (volts)	Current (milliamps)	Power (milliwatts)	Milliwatts per square centimeter
3.195 (diameter)	7.9	0.32	140	44.8	5.7
2.86 (diameter)	4.75	.40	110	44.0	9.25
1 × 2 (rectangular)	1.80	.40	42	16.8	9.35
0.5 × 2 (rectangular)	.80	.40	20	8.0	10.00

We mentioned that solar energy can be used to heat a working fluid. A possible system of this sort is shown in figure 1. Here a half-silvered, inflated Mylar plastic sphere (about 1 mil thick), 8.5 feet in radius, might serve as a collector,[13] at a weight cost of only about 8 pounds per 30 kilowatts of collected thermal energy.[14] An installation of this size would require roughly 100 square feet of radiator to reject waste heat (assuming a 10-percent overall conversion efficiency). This system is similar to a solar cell system in weight for a given power capacity. Again, meteorite effects represent an unknown, since they might puncture the collecting sphere or the

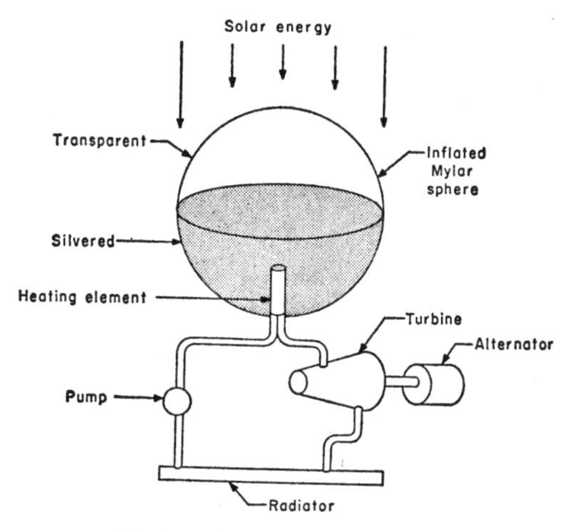

FIG. 1. Solar-powered alternator unit

radiator. The greater the actual meteorite hazard, the thicker and heavier the radiator would have to become. The development potential of solar energy sources seems good.

D. Nuclear Sources

Nuclear power sources can be available either with a reactor or through the use of isotopes.

Table 3 lists the specific power output available from a few selected pure and compound isotopes. These figures refer to newly produced isotopes; as time goes on, the power available from a given quantity will decrease. Because it is difficult to obtain pure isotopes, figures are also given for realizable compounds (oxides). Isotopes are generally quite expensive and may not be realistically suitable for really large-scale operations. However, isotopes prove interesting as a power source for satellites.

One direct way of using isotopes is in the form of a battery, through collection of beta particles (electrons) thrown off by an isotope such as strontium 90 (see figure 2). This will pro-

TABLE 3.—*Specific power of pure and compound isotopes*

Isotope	Specific power of pure isotope (watts per gram)	Specific power of attainable isotope compound (watts per gram)	Half-life	Source
Strontium 90	0.921	0.093	28 years	Fission product
Prometheum	.345	.053	2.6 years	Do
Polonium		141	138 days	Neutron irradiation of bismuth

NOTE.—Estimated fission-product power from nuclear power industry:

	Cumulative installed reactor megawatts	Approximate total beta and gamma kilowatts
Year:		
1965	3,600	2,000
1975	84,000	50,000

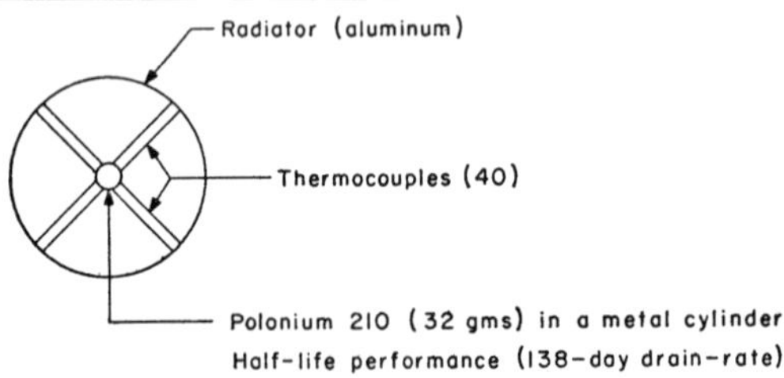

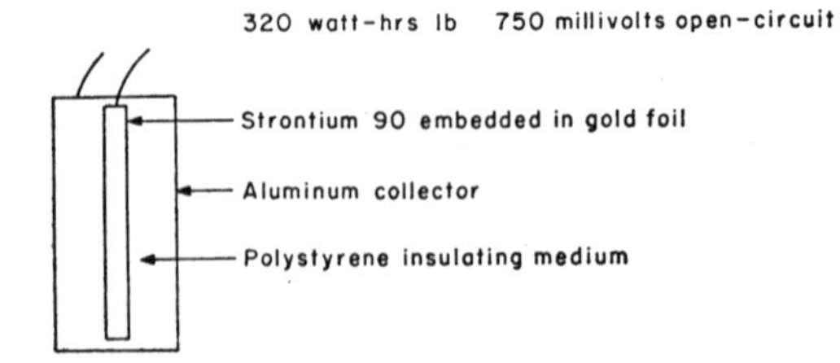

FIG. 2. Use of isotope power

duce a battery yielding several kilovolts, a gain over electrochemical batteries which are essentially low-voltage devices. However, while strontium batteries have a very long lifetime, they yield a rather low number of watt-hours per pound. Also, nothing can be done to alter the rate at which the isotope releases energy.

Another possibility might be to use an isotope, such as polonium, as a heat source and connect thermocouples from this source to a radiator. Such assemblies have yielded about 300 watt-hours per pound over their half-life. It is fairly easy to shield polonium to avoid the hazard of radiation. (Conventional metallic thermocouples are very inefficient; semiconductor varieties are better, but may be subject to radiation damage, etc.[15]) Isotope batteries are expensive: $375,000 for the one illustrated.[16]

Figure 3 shows the Elgin-Kidde prometheum battery (originally designed to operate a watch). All such batteries are inefficient in converting isotope decay to energy.

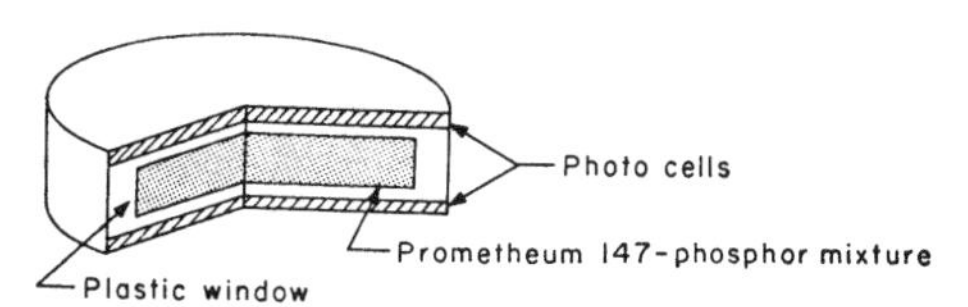

Half-life performance (25yr drain-rate)

Unshielded (90 r/hr) { 320 watt-hrs/lb; 22.8 watt-hrs/cubic in

Delivers 20×10^{-6} amps at 1 volt

Shielded (Tungsten alloy) { 8.5 watt-hrs/lb; 4.6 watt-hrs/cubic in

(Continuous exposure of human extremities permissible)

FIG. 3. Prometheum battery

Another possibility would be to use a polonium (or cerium) heat source as a boiler. Rotating conversion equipment could then probably provide higher conversion efficiencies, but with possibly less reliability.

In summary, while isotope power sources can probably outperform the electrochemical systems, they still have quite a few limitations. They dissipate energy at an unalterable rate, determined by the half-life of the isotope (290 days for cerium

144), and so must be designed for the expected peak demand. There is also the problem of throwing away excess heat at the start. Not only are they expensive, but there will probably never be enough material to permit large-scale use. Also, the hazard is quite high for a number of isotopes. (For example, polonium is a bone-seeker.) Furthermore, this hazard is greatest just at launching, when the unit is fresh. An isotope power source using cerium 144, called Snap I, is in development. The Martin Co. is the prime contractor, with Thompson Products as subcontractor for the rotating conversion equipment.[17]

Nuclear reactors have a criticality requirement which sets a lower limit to weight, and they may require more shielding. However, they do offer an almost unlimited lifetime and less hazard at takeoff. Reactors can generate large quantities of heat energy for very long times. For some time to come most of the weight of these power sources will be in the machinery needed to convert reactor heat to electricity. A reactor power supply called Snap II is also in development. The prime contractor is the Atomics International Division of North American Aviation. It is to use the same conversion equipment as Snap I.[18]

As previously stated, a polonium heat source can be used to heat a thermocouple to generate electricity. In a recently announced development, polonium encapsulated and attached to one end of a thermocouple heated the assembly to a temperature of about 1500° F. The unheated end of the thermocouple will radiate waste heat to space. It is estimated that such units, in development as Snap III, will deliver 3 watts for 6 months with a total weight of only 10 pounds. The cost of such a unit is expected to be about $25,000. The efficiency of the conversion from thermal to electrical energy is about 8 percent. No shielding from radiation would be required for most applications.[19]

Notes

[1] Crain, C. M., and R. T. Gabler, Communication in Space Operations, The RAND Corp., Paper P-1394, February 24, 1958.

[2] Research in Outer Space, Science, vol. 127, No. 3302, April 11, 1958, p. 793.

[3] Ingram, W. T., Orientation of Research Needs Associated with Environment of Closed Spaces, Proceedings of the American Astronautical Society, Fourth Annual Meeting, January 1958.

[4] Smith, J. G., Transistor Inverters Supply 400-Cycle Power to Aircraft and Missiles, Aviation Age, vol. 28, No. 6, December 1957, p. 112.

[5] Potter, E. C., Electrochemistry, Principles and Applications, Macmillan, 1956.

[6] Hydrox Fuel Cells—An Electrical Power Source for Civilian and Military Applications, Patterson-Moos Research Division of Universal Winding Co., Jamaica, N. Y.

[7] Miller, O. E., J. H. McLeod, and W. T. Sherwood, Thin Sheet Plastic Fresnel Lenses of High Aperture, Journal of the Optical Society of America, vol. 41, No. 11, November 1951, pp. 807-815.

[8] Huth, J. H., J. S. Thompson, and M. E. VanValkenburg, High Speed Impact Phenomena, Journal of Applied Mechanics, vol. 24, No. 1, March 1957, pp. 65-69.

[9] Proceedings of the 11th Annual Battery Research and Development Conference, Power Sources Division, U. S. Army Signal Corps Engineering.

[10] Jackson, E. P., Areas for Improvement of the Semiconductor Solar Energy Converter, presented at the Conference on Solar Energy at the University of Arizona, October 1955.

[11] Halsted, R. E., Temperature Consideration in Solar Battery Development, Journal of Applied Physics, vol. 28, No. 10, October 1957, p. 1131.

[12] Vavilov, V. S., V. M. Malovetskaya, C. N. Galkin, and A. P. Landsman, Silicon Solar Batteries as Electric Power Sources for Artificial Earth Satellites, Uspekki Fizicheskikh Nauk, vol. 63, No. 1a, September 1957, pp. 123-129.

[13] Such a collector is capable of producing considerably higher temperatures than can be used by existing materials.

[14] Ehricke, K. A., The Solar-Powered Space Ship, American Rocket Society, Paper No. 310-56, June 1956.

[15] Goldsmith, H. J., Use of Semiconductors in Thermoelectric Generators, Research, May 1955, p. 172.

[16] Hammer, W. J., Modern Batteries, Institute of Radio Engineers Transactions on Component Parts, vol. CP-4, No. 3, September 1957, pp. 86-96.

[17] Outer Space Propulsion by Nuclear Energy, hearings before subcommittees of the Joint Committee on Atomic Energy, Congress of the United States, 85th Cong., 2d sess., January 22, 23, and February 6, 1958; Col. J. L. Armstrong, p. 122.

[18] See footnote 17.

[19] See footnote 17.

8

Structures and Materials

The materials used in the construction of rocket boosters and space vehicles range from refractory high-density material for heat resistance to high-strength, lightweight materials to carry flight loads. For each application, the requirement for minimum weight is dominant. Any unnecessary pound of material used in the construction of the flight vehicles reduces the useful payload by at least a pound.

A. Possibilities for Materials Improvements

Some possible future improvements in structural materials are discussed in the following paragraphs.

STRUCTURAL STRENGTH

At present, designers have achieved structural configurations which use more than two-thirds of the maximum possible strength of the material. Further gains can be expected from novel designs, closer control of material properties and manufacturing tolerances, and the use of very large single shapes. Today's most efficient structural materials for use at normal temperatures, such as aluminum and titanium, can be surpassed soon by new materials, such as beryllium, and composite materials using high-strength filaments[1] or films.[2]

EXTREME THERMAL ENVIRONMENT[3]

The best current high-temperature metals, e.g., nickel and ferrous alloys, may soon be replaced by columbium. A better future prospect for higher temperatures is tungsten; however, more work needs to be done on tungsten to cope with such problems as its affinity for oxygen, before it will be practical. Another excellent prospect for high-temperature use is carbon, possessing a host of attractive properties. But structural use of carbon is currently restricted by its brittleness and the need for protecting it against oxidation, hydrogenation, and nitrogenation.

Ceramics such as carbides and oxides have very high melting points and show much promise for high-temperature use. They are not ductile, however, except in a few rare cases and under meticulously controlled surface conditions; but ceramics offer an attractive field for investigation.

Some materials possess outstandingly good properties but at the same time have certain prohibitive shortcomings (e.g., ceramics have excellent high-temperature strength but are extremely brittle). An ideal material would be a composite of two or more of these materials, each component being used for its outstanding contribution of a necessary property.

In some applications it may be advantageous to protect conventional structures from severe heat rather than to make a structure of high-temperature materials. Relatively brief encounters with a hot environment can be survived by the protection method, as in the insulation of rocket nozzles and re-entry nose cones.

B. Thin Shell Structures

In attempting to reduce weight, vehicle skins must be as thin as possible. But since very thin sections of material possess appreciable strength only under tension (stretching) loads and in no other direction, unique design and handling problems are created that will continue to require study and experimentation.[4]

There are several ways to use thin-walled structures. In some cases the structure can be internally pressurized to keep the walls from buckling. The net stretching force due to internal pressure is made greater than the compressional force due to flight loads, so that the tank walls experience no compression, and buckling is avoided.

Another method of stabilizing thin sheets is commonly used in the construction of conventional aircraft (figure 1). In this case (sheet and stringer construction), stiffening members (stringers) are fastened to the skin in the direction of the compressive load.

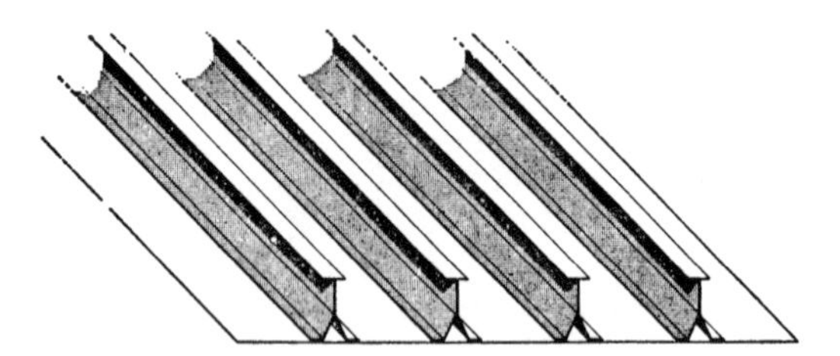

FIG. 1. Sheet and stringer construction

The same results can be obtained in a single piece by chemical milling or machining of a solid sheet to remove all metal except ribs or "waffles" that act as stringers (figure 2).

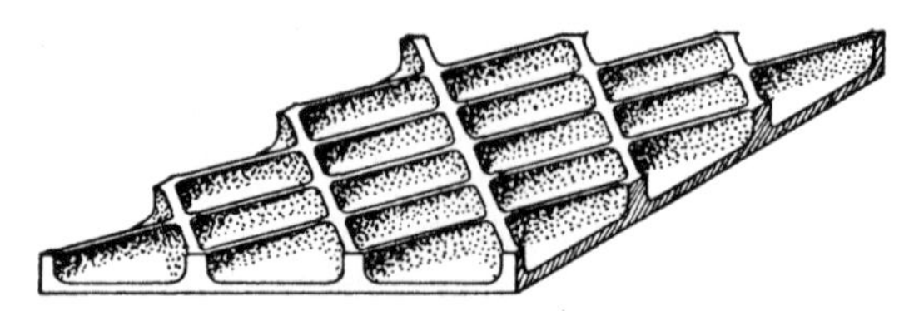

FIG. 2. Waffle construction

Thin sheets can also be stabilized against buckling by placing a lightweight supporting core between two sheets to form a "sandwich." The core might be in the form of honeycomb (figure 3) or corrugation, or possibly a light plastic or metal foam. Sandwich construction is becoming increasingly important in higher-temperature applications.[5] This type of construction should be useful in many types of flight vehicles.

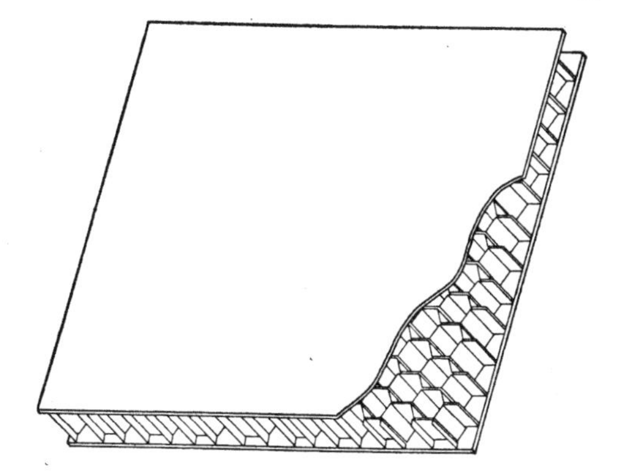

FIG. 3. Typical sandwich panel with hexagonal cell core

C. Large Area Structures

A number of space-flight undertakings, such as the collection of solar radiation for power generation, involve very large structures—covering areas of millions of square feet in extreme cases. The weight of these structures must be kept very low in order to make the associated systems feasible at all. Added to the problems of operating such structures in space are the problems (perhaps more serious) of packaging large amounts of fragile material compactly for carriage on launching rockets.

D. Structural Dynamics Problems

The bending and vibration of very light rocket structures interact with the flight-control system to such an extent that the structure and control system must be approached as an integrated design problem.

E. Temperature Control

The equilibrium temperature of a space vehicle is determined principally by the nature of the structural surface.[6] The radiation properties of this surface determine the relative rates of absorption of solar energy by the vehicle and the radiation of vehicle heat into space. This balance, along with the quantity of heat internally generated, determines vehicle temperature. Measures available to adjust this temperature balance include choice of surface color and smoothness of finish, as on Vanguard, Explorer, and Pioneer.[7] The overall surface characteristics can also be controlled in flight by operating "flaps" that cover or expose more or less surface area finished in black, as was done on Sputnik III.[8]

F. Meteorite Hazard

Data on entry of meteorites into the Earth's atmosphere can be used to estimate crudely how many meteorites a space vehicle will encounter by comparing the surface area of the Earth with that of the vehicle. A vehicle close to the Earth, such as a satellite, would be somewhat sheltered by the Earth from meteoroid collision. However, this reduction (by about one half) is small compared with other uncertainties—estimates of numbers of meteorites likely to be met vary by factors of 1,000 or more.

We can only guess at the number and size of meteorites, their speeds, and what effect their impacts will have on space vehicles. A great deal remains to be done before we understand the phenomenology of high-speed impacts, and there are no facilities yet available to experiment with particles moving at such high velocities.

As a specific example, consider a spherical vehicle with a diameter of 1 yard. Calculations based on available information indicate that the average period between punctures would be somewhere between 3 months and 170 years, if the skin were 1 millimeter thick. If the skin were one-quarter inch thick, the mean interval between punctures is variously estimated at 300 to 150,000 years.

Encounters with smaller particles (i.e., too small to penetrate) would be more frequent, of course. Thus, one might expect a "sandblasting" on the surface before penetration occurs. It has been estimated that surface erosion by small particles would be comparable to that produced by solar-particle radiation and by interaction with the gases of the solar corona. It has been estimated that the erosion due to these three effects would destroy the optical properties of a surface after about one year.

G. Multipurpose Structures

Because of the great premium on weight reduction in space vehicles, designs that use a single item of structure for more than one purpose would be highly desirable. It has even been suggested that material for propellant tankage, say, be made of combustible material (like lithium, for example) that might itself be used as fuel.

H. Additional Areas of Investigation

Areas in which important uncertainties may exist include the effects on various materials of a prolonged stay in a total vacuum (particularly critical with respect to lubricants[9] and paints and containers for gases and fluids) and prolonged exposure to radiations.

Notes

[1] Hoffman, G. A., A Look at "Whiskers," Astronautics, August 1958, pp. 31-33, 68.

[2] Hoffman, G. A., Beryllium as an Aircraft Structural Material, Aeronautical Engineering Review, February 1957, pp. 50-55.

[3] Hoffman, G. A., Materials for Space Flight, paper presented at Space Exploration Meeting, San Diego, California, August 5, 1958.

[4] Sandorff, P. E., Structures for Spacecraft, American Rocket Society Paper No. 733-58, November 20, 1958.

[5] Kaechele, L. E., Minimum-Weight Design of Sandwich Panels, The RAND Corp., Paper P-1071, October 16, 1958.

[6] Sandorff, P. E., and J. S. Prigge, Jr., Thermal Control in a Space Vehicle, Journal of Astronautics, spring 1956.
[7] Medaris, Maj. Gen. J. B., The Explorer Satellites and How We Launched Them, Army Information Digest, vol. 13, No. 10, October 1958, p. 5.
[8] The Third Soviet Artificial Earth Satellite, Pravda, May 18, 1958, p. 3.
[9] Freundlich, Martin M., and Arthur D. Robertson, Lubrication Problems in Space Vehicles, Airborne Instrument Laboratory, A Division of Cutler-Hammer, Inc. (c. 1958).

9

Flight Path and Orientation Control

A. Control in Powered Flight

The primary function of a control system during powered flight is to orient and stabilize the rocket vehicle. To turn the vehicle in some desired direction, it is necessary to develop *torque*. Control ceases and the body is stabilized when sensing instruments, usually gyroscopic devices, indicate that the proper attitude has been achieved.[1]

During the propulsion phase, control torque is usually developed either by aerodynamic forces acting on control surfaces or by rocket forces.

Techniques for aerodynamic control are essentially the same as those used in conventional aircraft. At very high altitudes, however, aerodynamic surfaces become ineffective because of low air density. Other means of producing torques are required in the vacuum environment of upper altitudes and space.

The method employed on the V-2 and Redstone missiles is to deflect the exhaust jet of the main propulsion motors by carbon vanes (figure 1). Both missiles actually use aerodynamic control surfaces, as well as jet vanes. At takeoff, the missile speed is zero, and the jet vanes provide stabilization. As the missile speed increases, the aerodynamic surfaces become effective and aid in the control process.[2]

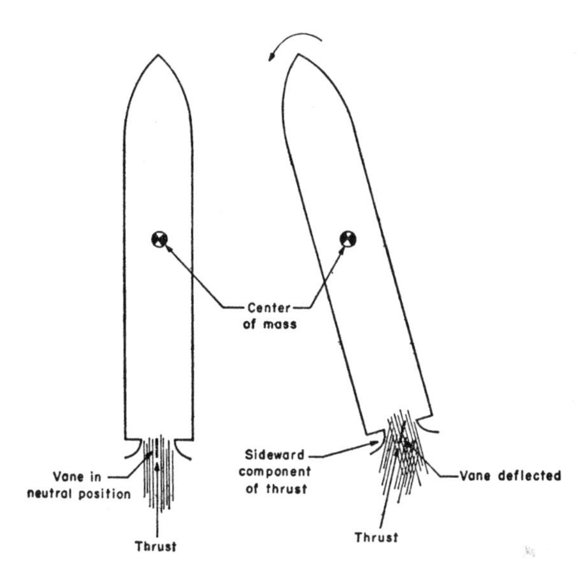

FIG. 1. Jet-vane control

A variation of the jet vane is the jetavator.[3] This device is a ring placed around the circumference of the motor nozzle. Deflecting the ring deflects the exhaust, just as with vanes. However, the jetavator has the advantage of not causing propulsion losses when in the neutral position, since, unlike vanes, it does not interfere with the exhaust flow. Jetavators have been most frequently applied to control of solid-propellant rockets.

Control of liquid rockets is frequently based on swiveling the entire rocket motor, as on the Thor.[4-6] Separate rocket nozzles may also be used to obtain a sideward component of thrust for control. Such small separate rockets are usually employed to control orientation about the longitudinal or *roll* axis of a vehicle.

Regardless of the mechanical details, the amount of control torque produced by the ejection of mass from the vehicle is equal to the sideward reaction force multiplied by the perpendicular distance from the center of gravity to the line of action of the sideward force.

By deflecting the rocket exhaust or by swiveling the rocket motor, the body of a vehicle may be rotated in any desired manner. Since the thrust from the main propulsion system is normally along the longitudinal axis of the craft, rotating the vehicle also rotates the direction of primary thrust in space. Adjusting the direction of the thrust in this manner may be used in turn to control the direction of flight of the vehicle.[7]

The magnitude of control torque required during the propulsion phase is determined by the various disturbing influences which act upon the vehicle during this period, including aerodynamic effects, wind, sloshing of propellants,[8] and various structural effects, such as body bending.[9]

B. Orientation Control in Free Flight

NATURE OF THE CONTROL PROBLEM

In unpowered flight in space the principal control problem is that of controlling the attitude or orientation of the vehicle with respect to a specified reference system. The techniques used in attitude control differ rather widely for various kinds of space missions.[10]

The attitude control problem has two major parts: (1) applying control torques, and (2) establishing an orientation reference system.

CONTROL TORQUES

An aircraft uses surfaces to deflect the airstream to obtain control torques. Without an external airstream to deflect, space vehicles are able to produce torques only through the use of self-contained mass. In some special cases there are exceptions to this rule, which will be considered later. In general, however, either a mass must be ejected by rocket units, or masses must be rotated internally to produce the required reaction torque.

Unlike the large rocket motors used for primary propulsion, the small rockets used only for control rotate the vehicle about its center of mass and do not influence the flight path. Generally, the thrust of such units is small, a matter of a few pounds, and they may be of the monopropellant type using a gas stored under high pressure or decomposing hydrogen peroxide.

Control torques may also be produced by rotating masses within the craft. The physical principle involved is the law of conservation of angular momentum. A familiar illustration of the consequences of this law is the manner in which the rotational speed of a swivel chair or a piano stool may be controlled by extending and retracting weights. If the stool is spun with the weights extended and the weights are then retracted to a position closer to the axis of rotation, the rate of spin of the stool will increase. If the weights are extended outward from their original position, the spin rate of the stool will decrease.

Now consider a wheel mounted so that its spin axle is perpendicular to the longitudinal axis of the space vehicle (figure 2). If, initially, the wheel and the vehicle are not rotating, the angular momentum of the system comprising the vehicle and the wheel is zero. Now if the wheel is caused to rotate, it will have an angular momentum. However, the total angular momentum of the vehicle and the wheel must remain zero. Thus, the reaction of the vehicle to the motion of the wheel is to rotate in the opposite direction so that the total angular momentum of the system is zero.[11]

In a typical installation, three wheels with their spin axes at right angles to one another are usually required to control the vehicle about its roll, pitch, and yaw axes; rotating masses are used in a variety of ways.

One axis of the vehicle can be made to remain fixed in space by spinning the entire vehicle about the axis to be stabilized.[12] In spin-stabilizing the payload stage of a Moon rocket, as in the Pioneer lunar probe, the vehicle is spun at propulsion cutoff in an orientation which, when the vehicle reaches the Moon's vicinity, enables it to be placed in orbit around the Moon by the firing of a reverse-thrust rocket.

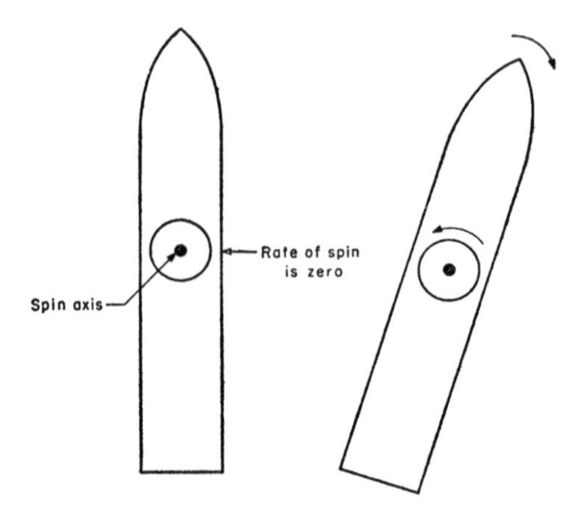

FIG. 2. Control by rotating mass

DISTURBING TORQUES

The attitude control system must be able to overcome any external or internal disturbing influences. Disturbing torques that must be considered when designing a stabilization and control system may be due to[13-17]

Residual aerodynamic forces.
Magnetic or electric fields.
Gravitational field gradients.
Solar radiation pressure.
Meteor impact.
The unintentional or uncompensated motion of internal masses.

The relative importance of these disturbing torques depends upon the particular vehicle under consideration.

Torques due to aerodynamic effects are of primary importance to satellites. With proper aerodynamic design, such forces might actually be used to aid in the attitude stabilization of a low-altitude, short-duration satellite.[18] At altitudes of 1 million feet and above, the air density is so low that aerodynamic effects usually are of little importance.

Torques due to magnetic fields might arise from induced currents in the conducting parts of the vehicle. Torques of this nature are also primarily of concern to satellites. Care in design details will assure that these effects are negligible.[19]

The Earth's gravitational field acts more strongly on a nearer mass than on a farther one. As a consequence, a torque arises that tends to force an elongated vehicle to orient itself in such a way that its long axis points toward the Earth's center. This *gravitational gradient* torque can also be used as a control torque if the vehicle is properly designed.[20-22]

The gradients of the gravitational fields of the other members of the solar system will also apply torques to space vehicles. However, unless the craft is in the vicinity of a planet, all such torques are negligible except that exerted by the Sun.

Torques due to solar radiation pressure[23] are present when there is an inequality in the effective reflecting area of the vehicle about the center of mass. Conceivably, such torques could be used for control purposes by adjusting the reflectivity of the proper part of the vehicle. At best, however, radiation pressure torques are quite small, particularly if care is taken to balance the effective reflecting area on either side of the craft's center of mass.

The importance of torques due to meteorite impact is rather difficult to determine. It is true that the impact of even a small meteorite could cause an angular disturbance; but the probability of such an occurrence seems to be very low.[24]

The importance of disturbances due to the motion of internal masses depends upon the detailed design of the vehicle. Unless great care is taken, the torques caused by internal rotating machinery, such as power units, or by people, are likely to be the predominant disturbing influence that the control system must handle.[25]

When a properly designed aircraft is given a small angular displacement, say in pitch, it will return to a neutral position without the need of control surface deflections. However, space vehicles requiring attitude stabilization must be under continuous active control. If a disturbance acts on the vehicle, only by applying a control torque can the resulting displacement be removed. Without control, even very small disturbing torques will eventually induce large orientation errors.

The very minuteness of the forces involved in space vehicle orientation control leads to a need for equipments that may be rather difficult to test under conditions at the Earth's surface.[26]

REFERENCE SYSTEM INSTRUMENTATION

The establishment of reference coordinates is a problem that is not unique to space vehicles. However, the great distances and long times over which the reference system of a space vehicle must be maintained lead to difficulties that do not occur in the case of aircraft.

Gyroscopes, which are merely elaborate spinning wheels, will maintain a fixed direction in space if they are not subjected to disturbances. In a practical case, however, it is just a question of time before the orientation of a gyro drifts away from the desired reference direction. The disturbing effects of high accelerations during launching, coupled with the relatively long duration of space flights, make it very difficult to establish a satisfactory preset reference system using gyros alone.

Instruments such as accelerometers or pendulums are of no use here. An accelerometer will always indicate zero in the weightless environment of a space vehicle, and a pendulum will simply assume a random position.

The stars provide a natural reference system which may be used in orienting a space vehicle. Star-tracking telescopes attached to the space vehicle can detect any change in the attitude of the vehicle and activate an appropriate torque-producing mechanism for correction.

Another device for instrumenting the orientation control of a satellite would be an optical one to scan the Earth's horizon, and thus establish the vertical direction.

The Earth's magnetic field can also be used to sense direction in space for orientation reference, as was done in Sputnik III.[27]

C. Vehicle Packaging

Vehicle shape and the distribution of internal equipments are of vital importance in orientation control. Two cases of particular importance are (*a*) spin stabilization and (*b*) satellite orientation.

For successful spin stabilization it is necessary that the axis of spin be the axis about which the vehicle has either maximum or minimum inertia,[28, 29] the axis of maximum inertia being the most stable. The third axis of a vehicle, having intermediate inertia, is not usable for stable spin. Thus a spin-stabilized vehicle should be either squat or long and thin. Uniform spherical shapes are generally undesirable for spin stabilization.

Stable orientation of a nonspinning satellite is greatly aided by use of the gravitational gradient torque, and this torque is available only if the vehicle is packaged to be relatively long and thin with its long axis pointing toward the center of the Earth. Distribution of internal equipment must be arranged to accommodate, and be usable in, this position.

Notes

[1] Miller, J. A., Designing Flight Control Systems for Ballistic Missiles, Aviation Age, November 1957, p. 92.

[2] Redstone, Flight, May 23, 1958, p. 705.

[3] Subsonic Snark Adds Effectiveness to SAC Forces, Aviation Week, vol. 69, No. 11, September 15, 1958, p. 50.

[4] North American Aviation, Inc., Rocketdyne Division, press release No. NR-38.

[5] Atlas Propulsion System Background, North American Aviation, Rocketdyne Division, press release No. RS-4, March 10, 1958.

[6] Thor Propulsion System Background, North American Aviation, Rocketdyne Division, press release No. RS-5, June 1, 1958.

[7] Orr, J. A., Powered Flight, Navigation, spring 1958.

[8] Wang, C. J., and R. B. Reddy, Variational Solution of Fuel Sloshing Modes, Jet Propulsion, November 1958.

[9] Beharrell, J. L., and H. R. Friedrich, The Transfer Function of a Rocket-Type Guided Missile with Consideration of Its Structural Elasticity, Journal of Aeronautical Sciences, vol. 21, No. 7, July 1954.

[10] Roberson, R. E., Attitude Control: A Panel Discussion: Problems and Principles, Navigation, vol. 6, No. 1, spring 1958.

[11] Angle, E. E., Attitude Control: A Panel Discussion: Attitude Control Techniques, Navigation, vol. 6, No. 1, spring 1958.

[12] Buchheim, R. W., Lunar Instrument Carrier—Attitude Stabilization, The RAND Corp., Research Memorandum RM-1730, June 5, 1956.

[13] Roberson, R. E., Attitude Control of a Satellite Vehicle—An Outline of the Problems, presented at the Eighth International Congress of Astronautics, Barcelona, 1957, American Rocket Society Paper 485-57.

[14] Vinti, J. P., Theory of the Spin of a Conducting Satellite in the Magnetic Field of the Earth, Ballistics Research Laboratory Rept. No. 1820, Aberdeen Proving Ground, July 1957.

[15] Roberson, R. E., Gravitational Torque on a Satellite Vehicle, Journal of the Franklin Institute, vol. 265, No. 1, January 1958.

[16] Manring and Dubin, IGY World Data Center, Satellite Rept. No. 3, May 1, 1958.

[17] Roberson, R. E., Torques on a Satellite Vehicle from Internal Moving Parts, Journal of Applied Mechanics, vol. 25, No. 2, June 1958.

[18] DeBra, D. B., The Effect of Aerodynamic Forces on Satellite Attitude, Lockheed Aircraft Corp., Missile Systems Division, Rept. No. MSD 5140.

[19] See footnote 14.

[20] Baker, R. M. L., Attitude Control: A Panel Discussion: Passive Stability of a Satellite Vehicle, Navigation, vol. 6, No. 1, spring 1958.

[21] Klemperer, W. B., and R. M. Baker, Jr., Satellite Librations, Astronautica Acta, vol. III, 1957.

[22] The Investigation of the Passive Stability of a Satellite Vehicle, Aeronutronic Systems, Inc., publication No. U-225, July 3, 1958.

[23] See footnote 12.

[24] See footnote 16.

[25] See footnote 17.

[26] Space Vehicle Attitude Control Experiments, Aeronutronic Systems, Inc., publication No. U-142, January 28, 1958.

[27] Stockwell, R. E., Sputnik III's Guidance System, Missile Design and Development, vol. 4, No. 9, September 1958, p. 12.

[28] See footnote 12.

[29] Bracewell, R. N., and O. K. Garriott, Rotation of Artificial Earth Satellites, Nature, vol. 182, September 20, 1958, p. 760.

10

Guidance

A. The Guidance of Space Vehicles

THE GUIDANCE PROCESS

The guidance process consists of measuring vehicle position and velocity, computing control actions necessary to properly adjust position and velocity, and delivering suitable adjustment commands to the vehicle's control system.

GUIDANCE PHASES

Guidance operations may occur in the *initial, midcourse, or terminal* phases of flight.

Ballistic missiles are commonly guided only during the initial flight phase, while rocket engines are burning.

A cruise type of missile, such as the Snark or Matador, uses midcourse guidance, operating continuously during cruising flight. Air-to-air missiles, such as Sidewinder, employ terminal guidance systems that lead the missile directly to the target on the basis of measurements on the target itself.

Any or all of these three kinds of guidance will be necessary for space flight, depending on the type of vehicle and mission involved.

INITIAL GUIDANCE

Space-flight missions in the immediate future will use ballistic rockets, and the guidance of such vehicles will be extensions of current ballistic missile guidance techniques. For ballistic missiles, guidance (sometime referred to as *ascent guidance*) is used only during powered flight, and guidance accuracy depends upon accurate establishment of flight conditions at the point of transition from powered to free flight. The necessary initial free-flight conditions are the position and velocity.

There is no single set of initial conditions required to arrive at a specified target, but rather there is an infinite number of possible free-flight paths originating at points in space in the vicinity of some nominal starting point which terminate at the desired destination. For each such point there is a corresponding proper velocity. It is the task of the guidance system to cause the rocket to take up one of these free-flight paths. The path selected for a given flight is determined ad hoc by the system as a running decision process during powered flight, on the basis of a complex of criteria involving structural loading, propellant economy, accuracy, etc. This approach is used because of the extreme difficulty of guiding a rocket along a single fixed powered trajectory in the face of detailed uncertainties in engine performance, aerodynamic and wind effects, component weights, etc. While the number of acceptable trajectories is infinite, practical limits on rocket behavior actually restrict the usable range to a zone fairly close to a nominal "optimum" trajectory.

When a suitable combination of position and velocity is reached, the guidance system must immediately signal cutoff of the propulsion system—in fact, it must signal cutoff a little ahead of time on the basis of a prediction process to compensate for time lags in the operation of engine controls.

Whether the rocket is an ICBM, a satellite launcher, or a launcher for a ballistic space vehicle, the powered flight guidance process is similar in all cases. The only material difference lies in the velocity required.

MIDCOURSE GUIDANCE

Using modern components and techniques, initial guidance will be accurate enough in most instances for establishing Earth satellites, placing payloads on the Moon's surface, establishing satellite orbits around the Moon, and the like. However, for travel to the other planets of the solar system, or very precise lunar missions, *midcourse* and *terminal* guidance systems will probably be required. Midcourse guidance will certainly be required for vehicles using electrical propulsion systems for long periods in space.[1]

Some measurements and computations for midcourse guidance can be made at Earth stations when a vehicle is not too far away; however, the limited range capabilities of optical and radar observations severely circumscribe such operations. For flights to regions substantially farther than, say, the Moon, except perhaps for probing flights, a guidance capability in the vehicle will be needed.

The guidance process then involves: measurements on board the spacecraft by a navigator or automatic equipment, based on star or planet sightings; computations of suitable control actions; and signaling operation of control rockets.[2] If suitable instrumentation is available the guidance process might be continuous or nearly continuous with periodic small steering corrections. If a human navigator is involved, observations would probably be made periodically.

The form of the computation process will depend somewhat upon the kind of instruments used for observations; however, it will probably be some variation of the techniques used by astronomers to determine the orbits of celestial bodies. One large difference, of course, is the time available for determining a precise orbit. For space navigation this determination must be made promptly and reliably, while the astronomer may proceed in a leisurely manner with careful checking.

TERMINAL GUIDANCE

Terminal guidance systems by their nature require some form of information from the target, such as infrared radiation, a radar echo, etc. This intelligence is then used to steer the vehicle to its destination,[3] whether it be a planet, a satellite, or a military target.

INERTIAL GUIDANCE SYSTEMS

Basically, an inertial guidance system consists of three accelerometers mounted on a gyro-stabilized platform, and some form of computer.[4-7]

Accelerometers are small mechanical devices that respond to accelerations of the vehicle. Each accelerometer measures acceleration in a single direction; therefore the three accelerometers are used to take measurements of the complete motion of the vehicle in space. These instruments can be used only during powered flight—in the "weightless" environment of free space they indicate nothing, regardless of the vehicle's trajectory.

The stabilized platform isolates the accelerometers from rotational motions of the vehicle and maintains the proper orientation of accelerometer axes.

The computer operates mathematically on the accelerometer indications to determine the true position and velocity of the vehicle and to give steering commands to the control system.

The critical components are the stabilization gyros and the accelerometers.[8] Disturbances due to imperfections in the gyros cause the platform to "drift" or rotate slowly. This drift rotation causes misalignment of the accelerometers—the "up and down" measuring accelerometer begins to measure "left and right" motion, so to speak, with resulting errors in navigation.

Imperfections in the accelerometers also result in incorrect determinations of position and velocity. At present, accelerometer inaccuracy is the largest single source of error in ballistic missile inertial guidance systems.

Inertial guidance systems are based entirely on measurements of acceleration, and involve no contact with the world outside the vehicle after launching. Thus, there is no known way of interfering with their operation—a point of some importance in military applications.

RADAR GUIDANCE SYSTEMS FOR BALLISTIC ROCKETS

Radars, and other similar radio devices, are also commonly used to determine vehicle position and velocity.[9-11]

Radar guidance systems are principally distinguished by the configuration of antennas used and the method of determining velocity information.

All radar systems for guidance use one or more radar beacons (repeater transmitters) in the rocket, a ground computer, and a radio command link to the vehicle. The achievable accuracy is largely affected by the baseline length or distance between individual antennas on the ground, the path of the vehicle during propulsion, the unpredictable features of the propagation path between the radar system antenna and the vehicle, and, of course, equipment errors.

These guidance systems, like any radio device, are susceptible to jamming and other interfering measures. However, modern electronic design practice includes rather effective means for counteracting external interference.

COMBINED RADAR-INERTIAL SYSTEMS

It is possible, in principle, to combine the measurements of vehicle velocity made by a radar and by an accelerometer to obtain a better overall measure of velocity than can be obtained by either measurement technique alone.[12] A dual system is to be preferred because radar systems are better for measuring slowly changing velocities, while accelerometers are better for rapidly changing velocities, and both occur in rocket flight. This process, of course, complicates the guidance system, adding either ground and airborne radar components to an inertial system or an airborne inertial system to a radar guidance system.

ILLUSTRATIVE ACCURACY DATA

An ICBM at a range of 5,500 nautical miles will have an impact miss distance in the direction of flight of approximately 1 nautical mile for an error of 1 foot per second in the magnitude of its cutoff velocity. The following table gives representative miss distances (in nautical miles) calculated for 1-foot-per-second error in the magnitude of velocity at thrust cutoff for various space-flight missions:[13]

TABLE 1.—*Miss distances*

Destination:	Nautical miles
Point on Earth (5,500 nautical mile ICBM)	1
Moon	20-100
Mercury	40,000
Venus	25,000
Mars	20,000
Jupiter	65,000
Saturn	200,000
Uranus	700,000

For establishing a satellite at an altitude of 1,500 miles, for example, an error of 1 foot per second in orbital velocity will cause the orbit to depart from circularity by about 1 mile.

NAVIGATION IN SPACE

Each "fix" on the orbit of an unpowered vehicle in free space (under the gravitational attraction of a single body, the Sun) requires six independent measurements of vehicle position or velocity.

A number of combinations of quantities might conceivably be measured by optical or radar instruments to obtain the six independent measurements. Some possible examples are:

Two sets of radar range measurements, separated by a sufficient interval of time, made on three celestial bodies moving in precisely known orbits around the Sun.

Three photographs of a major planet against its background of stars, separated by sufficient intervals of time.[14]

Three radar range measurements and three radar velocity measurements made simultaneously, at a known instant of time, on three separate celestial bodies moving in precisely known orbits around the Sun.

These measurements must be related to a known reference system, and this reference system must be maintained by some form of instrumentation on board the rocket. The best known way of implementing such a reference system is by a gyro-stabilized platform. Three mutually perpendicular directions or axes precisely defined on the platform can be oriented so that these three directions are parallel to the directions defined by the axes of the reference system to be used in navigation. The gyros will act to keep the reference system from rotating in space no matter how the vehicle moves. Over a long time, however, imperfections in the gyros will cause the platform and the reference to drift out of alignment. To remove these errors it is necessary periodically to correct the platform alignment by sightings on the fixed stars.

If some rocket thrust is used (an ion rocket, for instance), the trajectory followed by the spacecraft will not be so simply determined as that of an unpowered vehicle, and more frequent measurements will be required for use in more elaborate guidance computations by automatic equipment.[15]

B. Physical Constants

Precision in guidance depends not only upon the performance of instruments, but also upon the accuracy of the fundamental standards used, directly or indirectly, in the guidance measurements.[16-20]

One of the most basic and obvious of these fundamental constants is the specification of distance between the starting point and intended target point of a desired trajectory. The preciseness of this distance is dependent mainly upon the accuracy of our knowledge of the Earth's radius. The current figure for the mean radius of the Earth is probably correct to within a few hundred feet. If the error is, say, 300 feet, the strike error at the end of a 5,500-nautical-mile ballistic trajectory will be a little less than 1,000 feet—there is a magnification factor of about 3 between radius error and impact error at such a range. The magnification factor increases sharply with increasing range.

Another basic constant is the mean value of gravity at sea level. The current estimate of "g" is probably correct to within less than 0.001 percent. This mean value of g, together with the mean radius of the Earth, effectively defines a quantity proportional to the total mass of the Earth. The total mass of the Earth figures in all trajectory calculations. For example, the period of a satellite is dependent upon this parameter; and, in fact, measurement of satellite periods is a good way to determine the total mass of a parent body. As a result, we have good data on the masses of those planets that have satellites (moons), and less accurate data on the masses of planets without moons.

Since the Earth is not a perfectly uniform sphere, it is not enough to specify its mean radius and gravity. It is also necessary to specify how much it departs from the ideal sphere. The principal one of these additional specifications is the "flattening," the measure of difference between the radius of the Earth at the poles and at the Equator. The Earth is about 30 miles thicker across the Equator than from pole to pole.

The net assembly of these geophysical constants is known with a precision that is adequate for most space-flight purposes except for flights to other planets without midcourse or terminal guidance. Present knowledge is also probably inadequate for circumlunar flights intended to return to a small predetermined recovery area on the Earth; but the errors here can also be made good by midcourse or terminal guidance.

The mass of the Moon is also a parameter of importance for lunar and interplanetary flights. The Moon's mass is roughly one-eightieth of the Earth's; and current estimates are probably correct to within about 0.3 percent. This precision is adequate for all but very refined lunar flights—particularly for the circumlunar case just mentioned.

The constant of greatest importance in interplanetary flights is the astronomical unit—the mean distance from Earth to Sun. All of the dimensions used in the solar system by the astronomer are known to very great precision using the astronomical unit as a reference base; for example, the distance from Earth to Mars at a given instant can be obtained very accurately, from available data, in astronomical units. However, the length of the astronomical unit in Earth units like meters or statute miles is rather poorly known; and, therefore, the distance from Earth to Mars is rather poorly known in terms of miles. No vehicle using initial guidance alone is likely to make a hit on any of the planets (except by accident) until the length of the astronomical unit is better known in terms of Earth standards usable in guidance systems. The expected miss in shots to Venus and Mars, from this source of error, is likely to be many tens of thousands of miles.

These physical constants of primary relevance to astronautics are part of the general family of astronomical constants, and any improvements in their precision will be reflected in improvements in others.[21]

Another related area that needs wholesale improvement is detailed specifications of the high upper atmosphere, which are vital to accurate determination or prediction of satellite orbits.

In the future, it will be necessary to have the same kind of refined knowledge about the other planets as is now needed for the Earth.

Notes

[1] Spitzer, L., Interplanetary Travel Between Satellite Orbits, Journal of the British Interplanetary Society, vol. 10, No. 6, November 1951.

[2] Press, S. J., An Application of Solar Radiation to Space Navigation, Douglas Aircraft Co., Inc., Engineering Paper No. 655.

[3] Gates, C. R., Terminal Guidance of a Lunar Probe, Jet Propulsion Laboratory, External Publication No. 506, California Institute of Technology, May 14, 1958.

[4] Wrigley, W., R. B. Woodbury, and J. Horvorka, Inertial Guidance, Institute of Aeronautical Sciences, Preprint No. 698, 1957.

[5] Newman, T., Inertial Navigation, Arma Engineering, vol. 2, No. 2, October-November 1958, p. 8.

[6] Bishop, L. E., and E. W. Tooker, Inertial Platforms: The Key to Modern Guidance, Arma Engineering, vol. 2, No. 1, July-August 1958, p. 4.

[7] Frye, W. E., Fundamentals of Inertial Guidance and Navigation, Lockheed Aircraft Corp., Missile Systems Division, Rept. No. LMSD-2204, August 5, 1957.

[8] Draper, C. S., W. Wrigley, and L. R. Grohe, The Floating Integrating Gyro and Its Application to Geometrical Stabilization Problems on Moving Bases, S. M. F. Fund Paper No. FF-13, Institute of Aeronautical Sciences, 1955.

[9] Campbell, J. P., Interim Engineering Report A18-1 on Doppler Velocity for Space Navigation, Wright Air Development Center Contract AF33(616)-5487, General Precision Laboratories, April 1958.

[10] Campbell, J. P., Interim Engineering Report A18-2 on Doppler Velocity for Space Navigation, Wright Air Development Center Contract AF33(616)-5487, General Precision Laboratories, June 1958.

[11] Campbell, J. P., Interim Engineering Report A18-3 on Doppler Velocity for Space Navigation, Wright Air Development Center Contract AF33(616)-5487, General Precision Laboratories, August 1958.

[12] Mundo, C. J., Aided Inertial Systems, Arma Engineering, vol. 2, No. 2, October-November 1958, p. 15.

[13] Ehricke, K. A., Instrumented Comets: Astronautics of Solar and Planetary Probes, Convair Rept. AZP-019, July 24, 1957.

[14] Porter, J. G., Navigation without Gravity, Journal of the British Interplanetary Society, vol. 13, No. 2, March 1954.

[15] Roberson, R. E., Remarks on the Guidance of Ion-Propelled Vehicles, North American Aviation, Autonetics Division, June 21, 1958.

[16] Herrick, Samuel, Formulae, Constants, Definitions, Notations for Geocentric Orbits, Systems Laboratories Corp., Rept. SN-1, May 28, 1957.

[17] Herrick, Samuel, R. M. L. Baker, and C. G. Hetton, Gravitational and Related Constants for Accurate Space Navigation, American Rocket Society, Preprint No. 497-57.

[18] Buchheim, R. W., Motion of a Small Body in Earth-Moon

Space, The RAND Corp., Research Memorandum RM-1726, June 4, 1956.

[19] Study of High-Precision Geocentric Orbits, Aeronutronics Systems, Inc., Doc. U-094, August 12, 1957.

[20] High-Precision Orbit Determination, Aeronutronics Systems, Inc., publication No. U-220, June 27, 1958.

[21] Clemence, G. M., On the System of Astronomical Constants, The Astronomical Journal, vol. 53, No. 6, May 1948.

11

Communication

Much has already been accomplished in space communications, as evidenced by the following:

> The *Minitrack*[1] and *Microlock*[2] systems and other communication equipment associated with our own satellite and lunar probe programs.
>
> The communication systems associated with the sputniks.
>
> The communication design studies associated with military uses of satellites.

It is probable that for a long time to come communication techniques in space will grow out of those techniques already in use in current space-flight programs, and out of closely related techniques developed in fields such as air defense and *radio astronomy*. Basically, the same equations will govern the propagation of electromagnetic energy and the transmission of information in space as on the Earth.

We may, of course, expect certain practical differences between the conditions of space-flight communications and those of terrestrial communications. Among these are the following:

> The space environment. This refers both to the physical environment in which equipment will operate and to the propagation environment of interplanetary space and the various atmospheres and ionospheres of extraterrestrial bodies.

Very large communication ranges.

Severe size and weight limitations in space vehicles and severe reliability requirements for unattended operation over long periods of time.

One can combine what is known about the conditions under which space communications will have to operate with the basic equations of communication theory to predict the general lines of research and development which will be needed to accomplish the communication tasks required by space-flight programs. The discovery of any really novel effect—such as an unexpected propagation effect in one of the planetary atmospheres, or the possibility of using some other means than electromagnetic energy in communications—must await further developments.

Communications engineers have already invesigated in considerable detail the basic requirements in terms of radiated power, antenna performance, and so forth, for a variety of space-communications tasks.

The factors of primary importance in communications are the following: radiated signal power; area and directivity of transmitting and receiving antennas; communication frequency; receiver sensitivity; external interference; communication range; and channel bandwidth. Others of importance are termed "loss factors"; these include absorption losses, polarization losses, losses due to inefficient conversion of consumed power to radiated power, and losses due to inefficient types of modulation.

The feasibility of communicating across lunar distances (about 240,000 miles) with present components is well established. With components that can be made available in a relatively few years, communication will be possible over distances as great as 50 million miles (about the distance to Mars or Venus when these planets are relatively close to the Earth) although the communication bandwidth for such a distance would probably be small. For communication to Jupiter (about 500 million miles) or farther, additional technical advances would be required.

The following are some areas of research and development which are important for communications tasks associated with space flight:

Electrical energy and power sources. Clearly, space communications will demand strong electrical-energy and power sources. This demand may limit the useful lifetime of a space vehicle, as in the case of the early Earth satellites.

Radio-frequency power sources. Research is necessary to improve the efficiency of converting energy from various sources to electromagnetic energy of the desired frequency.

Data storage and data encoding. Many situations will arise in space-flight applications where it is necessary to store information for retransmission at some later time, or at a slower rate than it is received. Also, greater communication efficiency, resulting in power savings, can be achieved by improving methods of signal encoding or modulation.

Receiver sensitivity. In communication engineering, interference with a received signal is referred to as "noise." There are two kinds: receiver noise, which arises in the receiver itself, and external noise. Major sources of external noise are man-made interference, and solar and cosmic radiations. At the present time, receiver noise, in the majority of cases, is the most important factor in the sensitivity of signal reception. However, certain ultrasensitive types of receivers, such as *cooled detectors* and *masers,*[3, 4] are well along in development, and these may reduce receiver noise to 10 percent or as little as 1 percent of that in present-day receivers. With such receivers, external noise would become the dominant type of interference in space communications, and because of it, overall interference levels might be reduced only 80 or 90 percent, even if receiver noise were reduced a hundredfold. Since power requirements vary with the overall interference level, however, even an 80-percent reduction is worth striving for.

Directive vehicle antennas. Antennas may be either omnidirectional—that is, radiating energy roughly equally in all directions or receiving energy equally from all directions—or directive, radiating to or receiving from a preferred direction. The power saving from using directive antennas is very large—from ten to a thousandfold, depending upon the directivity. However, the use of directive antennas on space vehicles would require a certain amount of "attitude stabilization," or pointing, of the antenna to insure that the energy is radiated to or received from the right direction in space. This could be achieved most easily by attitude stabilization of the vehicle itself. The better the attitude stabilization, the more highly directive all the antennas and the greater the resultant power savings. Attitude control of antenna beams on space vehicles, therefore, offers one of the most promising avenues for research and development.

Very large surface antennas. Radio astronomy, the air-defense surveillance net, and current space-flight tracking activities are building up a backlog of experience in the use of very large steerable ground antennas, which will be needed for many space communication missions.

Circuit components. Work must be continued on improved miniaturization and packaging of components; greatly improved reliability for unattended operation up to perhaps several years; and investigation of, and protection from, damage caused by meteoric impact or radiation in space.

Research on cosmic, solar, and other external noise sources. As noted above, solar and cosmic noise may become the main source of signal interference when ultrasensitive receivers come into operation. Radio astronomers[5] have already conducted much detailed investigation of the intensity of solar and cosmic noise with respect to direction in space, frequency, and solar activity. Continued research in this field is necessary with a view to compiling the most complete maps possible of noise intensity as related to these various factors.

Research on physics of the solar system. The fact that the Earth's gaseous atmosphere and ionosphere crucially affect present-day communication is well known. Similarly, the atmospheres and ionospheres of other bodies in the solar system will affect communication to or from the surface of these bodies. In the vicinity of the Earth's Moon, which has a negligible atmosphere, electron densities may reach the value of 1 million or 10 million per cubic centimeter, which would certainly affect communication in this region.[6] Even in the space between Earth and Moon, electron densities may be as much as 1,000 per cubic centimeter. This affects, among other things, the velocity of light in Earth-Moon space. More precise measurement of electron densities in these regions constitutes an immediate possibility for useful research.

Notes

[1] Easton, R. I., U. S. Naval Research Laboratory Report No. 5035, Project Vanguard Report No. 21: Minitrack Report No. 2, The Mark II Minitrack System, September 1957.

[2] Richter, H. L., W. F. Sampson, and R. Stevens, Microlock: A Minimum Weight Radio Instrumentation System for a Satellite, Vistas in Astronautics (proceedings of the first annual Air Force Office of Scientific Research Astronautics Symposium), Pergamon Press, 1958.

[3] Culver, W. H., The Maser: A Molecular Amplifier for Microwave Radiation, Science, vol. 126, No. 3278, October 1957.

[4] Higa, W. H., Maser Engineering, External Publication No. 381, Jet Propulsion Laboratory, California Institute of Technology, April 25, 1957.

[5] Proceedings of the Institute of Radio Engineers, radio astronomy issue, vol. 46, No. 1, January 1948.

[6] Chapman, S., Notes on the Solar Corona and the Terrestrial Ionosphere, Smithsonian Contributions to Astrophysics, vol. 2, 1957, pp. 1-14.

12

Observation and Tracking

A. Visual Observation

THE VISIBILITY OF SPACE OBJECTS

The apparent brightness and probability of detection of objects in space depend upon the following important factors, each of which must be specified in order to estimate the visibility accurately:

The object's physical characteristics: size, shape, surface texture, and color (if visible only by reflected light) or luminosity and composition (if visible by light that it emits itself).

The distance between object and observer.

The brightness and character of the background or field against which the object is being observed.

The sharpness of vision of the observer and the magnifying power of his optical aids (if any).

The absorption characteristics and the light-transmitting ability of the medium between object and observer.

The apparent motion of the object, and the change in its appearance with time (whether steady, flickering, flashing, etc.).

In normal experience, we are compelled to view extraterrestrial objects through the Earth's atmosphere, which scatters light, is often cloudy, and is in constant motion. The movement of the atmosphere, even in the clearest weather, imposes a severe limit on the resolving power of large telescopes (their ability to produce finely detailed images or photographs). Thus, a large telescope employed above the atmosphere—say, on the surface of the Moon—would have a greatly enhanced ability to resolve details on the surfaces of the planets and the Sun and would tremendously improve man's ability to explore the entire visible universe.

The factors influencing the visibility of objects in space, listed above, are quite obvious and well recognized in ordinary everyday experience. However, to take them all into account simultaneously in calculating the visibility of distant objects usually becomes quite involved. Since simple mathematical expressions which would be useful under a wide variety of conditions cannot be formulated, a number of examples are given below to illustrate some size-distance relationships in the detection of just barely visible objects.[1]

SPECIFIC CASES

Table 1 illustrates the fact that plane mirrors of even small size, reflecting the Sun's image, may appear to be as bright as quite large expanses of diffuse (hazy) white material. The mirror reflects the Sun's image in a narrow beam, however, so the direction of the mirror with respect to the plane of the Sun, the mirror, and the observer is very important; whereas, with a diffuse white object, these relationships are much less important.

As examples of light emitters:

An ordinary automobile headlight (if aimed properly) located on the dark half of the half-full Moon could be detected by the 200-inch telescope.

The detonation of several pounds of magnesium illuminant on the dark half of the half-full Moon could be detected with the aid of large telescopes from the Earth.

TABLE 1.—*Minimum diameters of objects for detection at the distance of the Moon as a point of light from Earth under good viewing conditions*

Location of object and optical aid employed	Type of object (visible by reflected sunlight): Diffuse white disc	Circular plane mirror
	Approximate diameters	
2,000 miles from full Moon:		
Naked eye	2,500 feet	14 feet
7 x 50 binoculars	380 feet	2.5 feet
10-inch telescope	47 feet	3.5 inches
100-inch telescope	4.7 feet	0.35 inch
On surface of full Moon:		
Naked eye	25 miles	720 feet
7 x 50 binoculars	3.5 miles	105 feet
10-inch telescope	800 feet	50 inches
100-inch telescope	80 feet	5 inches

To a human observer in space, at a distance from any planetary body, the most conspicuous object would, of course, be the Sun, which would appear to be somewhat brighter than it does when viewed through Earth's atmosphere. In directions away from the immediate vicinity of the Sun and its surrounding corona, the sky would appear to be black but more brilliantly star-studded than the night sky as seen from Earth under the most favorable conditions; the "twinkling" would be absent. All the familiar constellations would be visible at once, from Ursa Minor to Octans (the constellations above the North and South Poles of the Earth). Against this rich starry background, artificial objects would be difficult to detect unless made highly conspicuous by reason of color or brightness. Advanced search techniques would be needed to locate faint objects.

To an observer on the surface of the Moon facing the Earth, a similar sky would be presented, except that only one hemisphere could be observed at any one time, and in it the Earth would hang always in almost the same position above the horizon. The Earth would display phases (as the Moon does from Earth): about half its surface would be seen to be covered with clouds at all times; where not obscured by cloud cover, the oceans would appear dark, the continents lighter in color; the reflection of the Sun's image on the bodies of water would appear as a bright spot of light. At its brightest, the Earth would provide about a hundred times as much illumination as the full Moon as seen from Earth.

DETECTION OF EARTH SATELLITES

The problem of visually detecting satellites of the Earth from the surface of the Earth is especially complicated by the fact that small bodies visible only by reflected light must be in sunlight to be seen at all, and yet to the observer they must appear against a dark background. Thus it is only during the hours shortly after sunset and before sunrise that small satellites may be observed optically. Also, if they are close to the Earth, they may be seen only along a relatively narrow band on the Earth's surface beneath the track of the satellite. Larger satellites, of course, may be seen against more brightly illuminated sky backgrounds. As an extreme example, in order to be visible to the naked eye during daylight hours with clear skies, a satellite at an altitude of 1,000 miles would have to be about 600 to 700 feet in diameter (white sphere).

B. Infrared Observation

Infrared detection systems are playing an increasingly important role in scientific and military applications.[2-4] Two fundamental physical phenomena are responsible for this:

> Radiations in the infrared (heat) portion of the electromagnetic spectrum interact strongly with the molecular structure of matter.
>
> Every physical object emits thermal radiation, the radiation intensity increasing rapidly with its temperature.

Transmission or reflection of infrared radiation can serve as a probe of the structure and composition of chemical and biological matter. Not merely limited to laboratory analysis, infrared probes have permitted man to begin to understand the composition of the atmospheres of the planets and other celestial objects. The principal hindrance to such extraterrestrial observations has been the confusion offered by the intervening Earth's atmosphere. Observations from immediately outside the atmosphere will permit examination of planetary atmospheres with sufficient precision to study the environment that future planetary explorers will have to cope with.

Similarly, infrared observation of the Earth's atmosphere from an orbiting vehicle will permit measurements of cloud cover, water-vapor, and carbon-dioxide concentrations, and the like, which will be valuable for meteorological observations. In addition, infrared sensors in similar vehicles may have some usefulness for military reconnaissance and surveillance of the Earth. In particular, especially large sources of infrared energy such as afterburning engines of supersonic aircraft and exhaust flames of ballistic missiles may be detectable from satellites.

Infrared detection range is severely limited within our atmosphere by scattering and absorption, so that most military applications, such as guidance for air-to-air missiles of the Sidewinder type, are of a short-range (i.e., less than 10 miles) nature. However, the space environment will permit full exploitation of the sensitivity of infrared detectors, permitting much greater ranges. Of course, celestial sources will be detectable at even greater distances.

Infrared sensors will permit the surveillance of space vehicles in a manner distinct from either optical detection, based on reflected sunlight, or radar, based on reflection of radio waves. The usefulness of this technique will depend in part upon the amount of heat expended in future vehicle power supplies and propulsion systems. One advantage of infrared for ground-based observation of space objects is the ability of infrared to produce a clear image even through haze and scattered light.

Since a large portion of the radiant energies of sunlight lie in the infrared region, it is also possible to observe space vehicles with systems that combine both optical and infrared sensors. A detector such as lead sulfide is sensitive to both these classes of radiation, and may therefore be better in some applications than a narrow sensor limited to a single region of the spectrum.

Infrared can also be useful for detection of celestial objects in space navigation. For such an application, infrared possesses many of the advantages of optical techniques, such as directional accuracy and small equipment size and weight; however, it has one important and unique advantage due to the temperature-magnitude relationship inherent in celestial objects. The tremendous number of optically detectable objects may tend to cause confusion if an optical system is used for navigation or surveillance, unless some method for logical discrimination on the basis of space relationships is built in or is part of the skill of the human observer. A properly filtered infrared detection system, however, will be limited to detection of a much smaller number of celestial objects which, fortunately, include the most interesting near objects such as the planets, thus offering a simplified background problem. When combined with spectral analysis, infrared sensors would permit clear identification of a planet, for example, which may be of considerable value in navigation and terminal guidance of space vehicles.

C. Tracking

PURPOSE AND PROBLEM OF TRACKING

The general purpose of tracking is to establish the position-time history of a vehicle, for guidance, navigation, observation, or attack. The techniques employed are essentially the same for these various purposes. In current satellite and Moon rocket projects the relatively heavy components of guidance equipment are jettisoned after powered flight, and the vehicle is tracked in free flight by other radio or optical means. Navigation in space will undoubtedly require tracking of sources on the planets and in other vehicles as well as tracking of the vehicle itself from Earth and other bases.

TYPES OF TRACKING SYSTEMS

The principal types of tracking systems are:

Radar and radio systems.
Optical systems.
Infrared systems.

RADAR AND RADIO SYSTEMS

These systems employ radio frequencies that are generally in the range from 100 kilocycles to 30,000 megacycles. Below this frequency range, antennas with adequate directivity become too large to be practical, and ionospheric propagation difficulties become severe. Above this frequency range, there are, at present, practical limitations on the power that can be generated. There are also regions near the upper end of this frequency range (at least for Earth-based stations) that must be avoided because of water-vapor absorption and losses due to scattering by rain. In tracking against a background of cosmic noise certain frequencies and frequency regions must also be avoided.

Radar and radio systems may be further classified into active and passive systems, the first requiring transmitting equipment in the vehicle, generally referred to as a *beacon* or *transponder*. Passive systems depend upon the reflective properties of the vehicle to return the radio waves that hit it. These properties may be enhanced by the use of special reflectors or may be reduced by special surface treatment. Active systems are generally superior to passive systems with respect to range capability and tracking accuracy, but they require special equipment aboard the vehicle. Therefore, in general, active systems can be used only in connection with friendly vehicles in working order.

Radio tracking systems are also categorized as "continuous wave" and "pulse" systems depending upon the scheme used for measurement of range. Angle measurements are sometimes accomplished by a scanning technique in which the antenna pattern is moved either by mechanical or electronic means about the direction of strongest signal return. This method is employed by the more conventional types of radars and by some of the radio telescopes used in radio astronomy. Another method uses the principle of the *interferometer* to compare the phases of signals received in separate antennas on well-established baselines. This method is employed by the Minitrack [5] system used for the Vanguard satellite and the Microlock [6, 7] for the Explorer. The frequency of the returning signal from a vehicle being tracked depends not only upon the transmitted frequency but also on the relative motion of the vehicle and the tracker. This *doppler effect* makes it necessary to design the tracker to automatically follow the changing frequency. By the same token it is also possible to use this frequency change to measure the relative velocity of the vehicle and the tracker.

OPTICAL TRACKING SYSTEMS

Optical systems make use of the visible-light portion of the electromagnetic spectrum. They all consist essentially of a telescope mounted on gimbals to permit rotation about two axes. One type, the cinetheodolite produces a photographic record of the position of the target image with respect to cross hairs in the telescope, along with azimuth and elevation dial readings and a timing indication. With two or more such instruments on accurately surveyed baselines, the position of a target in space is obtained by simple triangulation. Tracking is usually manual or partially manual.

Another type of optical tracking instrument is the ballistic camera which determines angular position by photographing the vehicle against a star background. This instrument can be very accurate, but the data require special processing by skilled personnel and the time delay involved is sometimes a disadvantage. Schemes for making some, or all, of the procedures automatic are being considered. An instrument of the ballistic camera type designed especially for optical tracking of Earth satellites is the Baker-Nunn satellite-tracking camera.[8]

Angle tracking with optical equipment can be accomplished with much greater precision than with radio equipment, but, for Earth-based trackers, darkness, clouds, and haze limit the usefulness of optical equipment. Another limitation of optical trackers is the fact that, as we have seen, data reduction sometimes delays the output beyond the period of usefulness. The eventual solution is to make these procedures automatic.

INFRARED TRACKING

For tracking certain objects some advantage is gained by using the infrared portion of the spectrum. Infrared radiation from the object being tracked may provide a better contrast with the background radiation, and certain fog and haze conditions are more readily penetrated. In general, however, infrared radiation is also absorbed by the lower atmosphere, and the range and general usefulness of infrared tracking are enormously increased by carrying the equipment in high-altitude vehicles essentially above the Earth's atmosphere. The use of photoelectric detection and scanning techniques permits information concerning angular position to be "read out" automatically.

A rather new development in optical and infrared tracking is the use of television techniques to improve sensitivity, selectivity, and rapid readout characteristics of the tracker. Work in this field is being done both in this country[9] and abroad.[10] Essentially, a tracking telescope is fitted with the front end of an *image orthicon* (standard TV camera tube) followed by an *image intensifier* for amplification of photoelectrons. The results provide the ability to "chop off" the sky background and permit tracking in daylight as well as tracking of fainter objects at night.

D. Orbit Determination

Vehicle tracking data are raw material that must be processed mathematically to provide orbit information. Depending upon the application, this orbit information may be needed to establish accurately a vehicle's past behavior and its current position, in order to predict its position in the more-or-less distant future.

Preliminary orbit information is available to the party launching a space vehicle from prelaunch adjustments and from measurements made during the launching operation; and data concerning the orbit can be corrected and improved as further measurements are made. However, a party not privy to launching operations must be able to detect the vehicle and to determine its orbit from scratch, from observational data.

At present the major orbit computational centers in the United States are the Smithsonian Astrophysical Observatory in Cambridge, Mass.; the Vanguard Computing Center in Washington, D.C.; and Project Space Track of the Air Force Cambridge Research Center. These centers receive data from the various tracking stations around the world. The technique followed is first to derive a preliminary estimate of the orbit shape, then to refine and correct it as further data become available. Actually, two tasks exist for the centers responsible for orbit computation. One is to learn enough about the orbit quickly to be able to predict its positions for useful periods into the future, and the other is to derive in a more leisurely manner a "definitive orbit" giving the satellite's past history.

Prediction of satellite positions for days or weeks into the future is limited primarily by lack of knowledge of air-drag forces. For satellites with orbits which stay sufficiently high—e.g., the Vanguard satellite, 1958 Beta—predictions can be made several weeks in advance with errors of about a mile; while for a lower satellite—e.g., Explorer III—predictions of such accuracy can be made for only a day or so in the future. This situation should improve as information is gathered about the upper atmosphere, but there will always be uncertainty in predicting the drag force which will act on low satellites, particularly those with irregular shapes. If the position of a satellite is to be known to a fraction of a mile, then it will be necessary to continually revise the *orbital elements* to take care of unpredictable changes in the upper atmosphere.

If it is ever to be possible to predict satellite positions to within one-tenth of a mile or better, it will be necessary to improve the present knowledge of the distribution of mass in the Earth, and the density of air in the upper atmosphere. Satellites represent the best way of studying both; but provision must be made for numerous precise observations around the orbits and use must be made of satellites with precisely known shapes, preferably spherical.

Satellites stay on orbit for long periods—virtually forever if high enough. Thus, an accumulation of hundreds of satellites may be on orbit by the early 1960's as a result of various launchings for scientific, military, and perhaps even commercial purposes. Present satellite data-handling and computing methods will be unable to cope with the problems presented in identifying, cataloging, and keeping track of large numbers of satellites, and it is important that new methods be devised and implemented without delay.

E. The Location of Earth-Based Trackers

A number of factors—vehicle motion, Earth rotation, and the need for an unrestricted line of sight between the tracker and the vehicle—combine to dictate the number and location of Earth-based tracking stations for a particular application. Refraction, or bending, of radio and optical rays makes it possible to "see" objects below the horizon, but uncertainty in the refraction correction makes it necessary to restrict useful tracking data to elevation angles greater than zero. Refraction in the lower atmosphere is due to the presence of air molecules, water vapor, and other constituents, and the refraction of radio waves in the ionosphere is due to the induced motion of the electrons. The latter effect tends to decrease with increasing frequency of the radiation. Since water-vapor content, air densities, and electron densities are variable with time and place, refraction corrections can be made only approximately.[11] The refraction correction, as well as the uncertainty in the correction, is greater for low elevation angles. Therefore, useful tracking data can often be obtained only by avoiding low-elevation angles.

Another important factor affecting the selection of the number and location of Earth-based tracking sites is the effect of Earth rotation. For many applications, locations outside the United States are required to provide adequate coverage in longitude.[12] The choice of tracker locations also depends upon the particular application at hand, since this determines the precision required. For a radio tracker, zenith (directly overhead) passage yields the most accurate vehicle position and velocity information as well as the longest observation time. This is also true for optical trackers, although for existing optical tracking equipment the relative position of the Sun is a modifying factor.

For some vehicles, tracking is required merely to keep a record of their position, so that a vehicle may be picked up by a tracker when necessary for identification purposes. At the other extreme lie vehicles used for navigation purposes or for determining geophysical or astronomical data. Here, the number of trackers, their accuracy, and their location are all of great importance, and the tracking requirement for each application must be analyzed separately.

Notes

[1] Dole, S. H., Visual Detection of Light Sources On or Near the Moon, The RAND Corp., Research Memorandum RM-1900, May 24, 1957.

[2] Tozer, E., Uncle Sam's New Wonder Weapon, This Week magazine, November 16, 1958, p. 10.

[3] Estey, R. S., Infrared: New Uses for an Old Technique, Missiles and Rockets, June 1958.

[4] Powell, R. W., and W. M. Kauffman, Infrared Application to Guidance and Control, Aero-Space Engineering, May 1958.

[5] Mengel, John T., Tracking the Earth Satellite and Data Transmission by Radio, Proceedings of the Institute of Radio Engineers, vol. 44, No. 6, June 1956, p. 755.

[6] Sampson, William F., Henry L. Richter, Jr., and Robertson Stevens, Microlock: A Minimum Weight Radio Instrumentation System for a Satellite, Progress Rept. No. 20-308 ORDCIT project contract No. DA-04-495-ORD 18, Department of the Army, Ordnance Corps., Jet Propulsion Laboratory, California Institute of Technology, November 14, 1956.

[7] Sampson, William F., Microlock: Capabilities and Limitations, Technical Note No. HTR 58-009, Hallamore Electronics Co., November 4, 1958.

[8] Henize, Karl G., The Baker-Nunn Satellite-Tracking Camera, Sky and Telescope, vol. XVI, No. 3, January 1957.

[9] Gebel, Radames K. H., Daytime Detection of Celestial Bodies Using the Intensifier Image Orthicon, WADC Technical Note 58-324, October 1958.

[10] Report of the International Astronomical Union Meeting, August 1958, Moscow (to be published).

[11] Crain, C. M., Survey of Airborne Refractometer Measurements, Proceedings of the Institute of Radio Engineers, vol. 43, No. 10, October 10, 1955.

[12] Gabler, R. T., and H. R. O'Mara, Tracking and Communication for a Moon Rocket, Vistas in Astronautics, Pergamon Press, Inc., 1958.

13

Atmospheric Flight

A. Continuous Atmospheric Flight

OCCURRENCE OF AERODYNAMIC FLIGHT

Planetary atmospheres comprise a very small part of space; however, these atmospheres give rise to some very vital problems in space flight.[1] During exit from and entry into planetary atmospheres, hypersonic speeds (that is, speeds of more than about five times the speed of sound) will be characteristic of all space planetary vehicles. Thus, hypersonic aerodynamics will be involved in important phases of operation.

Strictly speaking, we should use the term "gas-dynamics" rather than "aerodynamics," since our consideration is not limited to the Earth's atmosphere. Furthermore, due to chemical effects, the "air" we are concerned with in hypersonic aerodynamics may be quite different from our customary ideas of air. However, we shall use "aerodynamics" as a term that does not preclude gases other than air.

LONG-TERM FLIGHT

The vehicle for sustained atmospheric flight at conditions of interest in astronautics is the hypersonic glide rocket.[2] Although not a true cruising-type vehicle, the glide rocket is much more closely related to the conventional airplane than is the ballistic rocket. It is a promising vehicle for manned hypersonic flight and for manned return from space. The glide rocket is boosted to initial speed and altitude by a rocket, much like a ballistic vehicle. It is then tilted over into a glide path and glides back to Earth, losing speed and altitude as it descends. To enhance accuracy and to avoid low impact speeds, the flight path of an unmanned glide bomber would probably end in a near-vertical dive. A typical flight path is shown in figure 1. Note that the vertical scale is considerably expanded, and that the entire flight path is within the atmosphere.

A likely intercontinental-range glide-rocket configuration is shown in figure 2. It is a long, thin, streamlined vehicle in the interest of high lift and low drag. The flat-bottom body and drooped nose provide the best *lift-drag ratio* for hypersonic flight.[3] The flares at either side of the body are essentially wings which increase the lift-drag ratio. The rate of aerodynamic heating along a glide path is low enough to permit the use of thin-skin construction except for hotspots near the nose and leading edges.

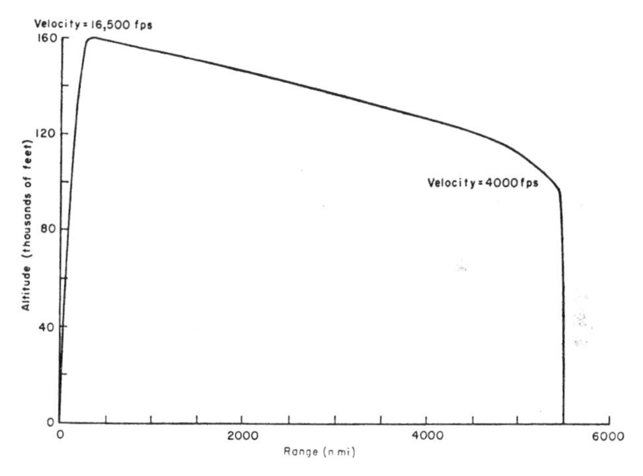

FIG. 1. Glide path

The greater efficiency of the intercontinental glide rocket, as compared with that of the ballistic rocket, is illustrated by the fact that for two vehicles of the same gross weight and payload the ballistic rocket will go only about one-third as far as the glide rocket, or, for the same range and payload, the glide rocket will weigh only one-third as much. If an ICBM were converted to a glide rocket of the same range and initial gross weight, the payload could be 8 to 10 times as great. There are, of course, many other factors that influence a choice between ballistic missiles and gliders.

A manned glide rocket involves essentially the same design considerations for the glide portion of its path, with the added requirement of a compartment suitable for human occupancy. However, the rocket ascent and the landing phase must be modified for human tolerance and safety. The takeoff acceleration of a typical unmanned glide rocket or ballistic rocket might start one-half g above normal gravity and increase to perhaps 10 g as the propellant is burned. For an inhabited vehicle, the maximum acceleration could be reduced to a tolerable limit—perhaps 4 g—by extending the time of powered flight, i.e., accelerating more gradually.

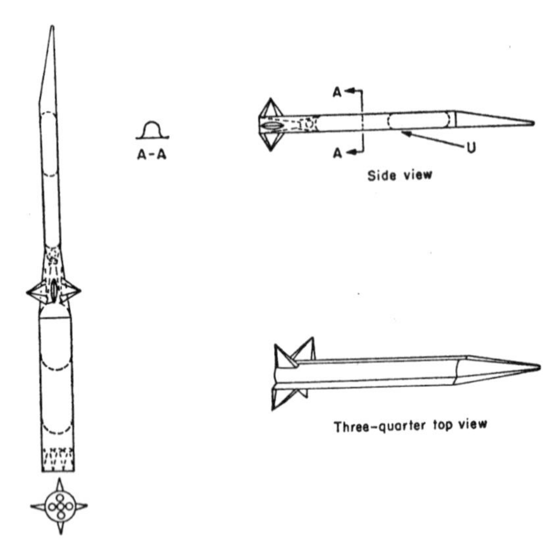

FIG. 2. Typical glide rocket

The final portion of a manned flight must involve essentially a conventional airplane type of landing or a parachute recovery, so that decelerations will not injure the human pilot. Ballistic re-entry, except for very shallow descents, involves decelerations well above human tolerance. Glide-path descents, however, are accompanied by very low decelerations—usually less than 1 g. While power-on landings appear to present no insurmountable problems, they do involve a weight penalty.

The glide rocket, which was introduced conceptually in this country nearly 10 years ago,[4] is closely related to the skip rocket which was first proposed by Sänger and Bredt.[5] The skip path is similar to that of a flat stone skipping over the surface of a pond; a lifting-type vehicle descends on a ballistic path from above the atmosphere; upon re-entry into the atmosphere, lift builds up with dynamic pressure, causing the vehicle to take an upward turn and be tossed from the atmosphere again. Thus, the skip path consists of a succession of ballistic trajectories each followed by a pullout.

From a pure flight-mechanics standpoint, the skip rocket is superior to both ballistic and glide vehicles. However, the increased structural weight resulting from the higher stresses and greater peak aerodynamic heating rates of the skip rocket reduce its net range below that of the glide rocket. Furthermore, the severe stresses associated with the most desirable skip path rule out its use for manned vehicles.

As indicated by these brief considerations, feasible hypersonic airplanes are primarily of the boost-glide type. It is expected that development of a hypersonic cruise aircraft must wait the development of an efficient sustaining powerplant, such as a hypersonic ramjet.

The initial ascent flight paths of spacecraft will be almost identical with those of ballistic and glide rockets.

The distinction between the supersonic and hypersonic flight regimes is not clear-cut; but, for most interesting vehicles, the dividing line falls close to Mach 5, five times the speed of sound. In hypersonic flow the shock wave lies close to the surface of the vehicle body; whereas, in supersonic flow the nose shock wave is fairly far from the body.

In the flight of a vehicle through the atmosphere, some of the body's kinetic (motion) energy is continuously being converted to thermal (heat) energy in the air, and some of this thermal energy is transferred to the body. The rate of conversion of kinetic energy to thermal energy and the rate of heat transfer to the vehicle surface increase approximately directly with air density and very sharply with increasing vehicle speed. Surface heating rates are thus most severe when high speeds occur at low altitudes, and can become more severe than any heating rates experienced in current technology.

Because air does not behave as a simple fluid under hypersonic flight conditions, hypersonic aerodynamics is much more complex than lower-speed aerodynamics. Unusual chemical and physical events occur in the violently heated air near a hypersonic vehicle. The high air temperatures cause excited molecular states, radiation, chemical reactions, ionization, and so forth, resulting in effects that further complicate the overall heat-transfer problem, and may also cause difficulties for radio transmissions to and from the vehicle.[6-8]

B. Atmospheric Penetration

TYPES OF ENTRIES

Severe problems of heating and stress confront any vehicle that must penetrate a planet's atmospheric mantle and survive to the surface. The cases of interest run from simple sounding rockets to manned vehicles returning from interplanetary trips. Several types of atmospheric entry paths are illustrated in figure 3.

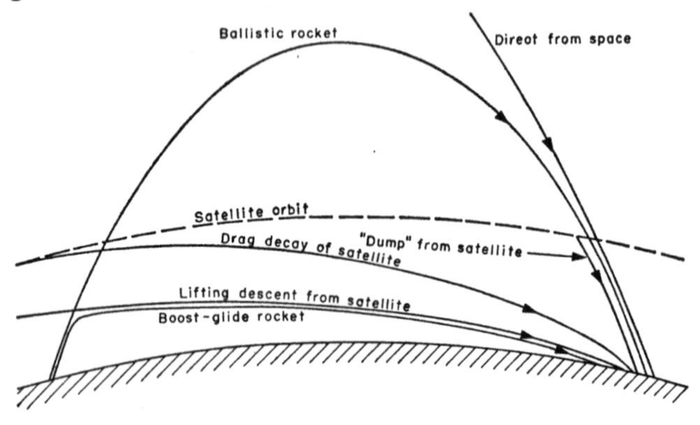

FIG. 3. Various types of atmospheric entry

It can be seen that the glide vehicle just discussed descends through the atmosphere in a more gradual fashion than the ballistic missile.

Descent from a satellite orbit may be accomplished either by waiting for the orbit to decay under the action of aerodynamic drag or by using rocket braking ("dump") to shift from the satellite orbit to a descent path. Depending upon the aerodynamic characteristics of the descending vehicle, the entry path may range from the gradual one of a glide vehicle to the steeper path of a ballistic vehicle.

A vehicle arriving from outer space will approach a planet with a velocity that is at least equal to the escape velocity characteristic of the planet. (In the vicinity of the Earth, the escape velocity is about 37,000 feet per second.) Several kinds of approach orbits of interest are illustrated in figure 4. A direct

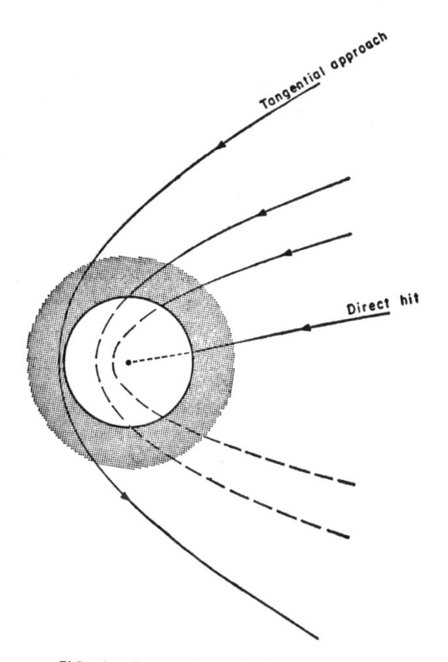

FIG. 4. Approach paths from space

hit on the planet would involve an entry path similar to that of a ballistic rocket, but with a higher velocity. A more gradual penetration can be accomplished by either approaching the planet tangentially (flatly) or by maneuvering into a satellite orbit before descending. Shifting into a satellite orbit can be accomplished either by rocket braking or by the aerodynamic braking procedure illustrated in figure 5.

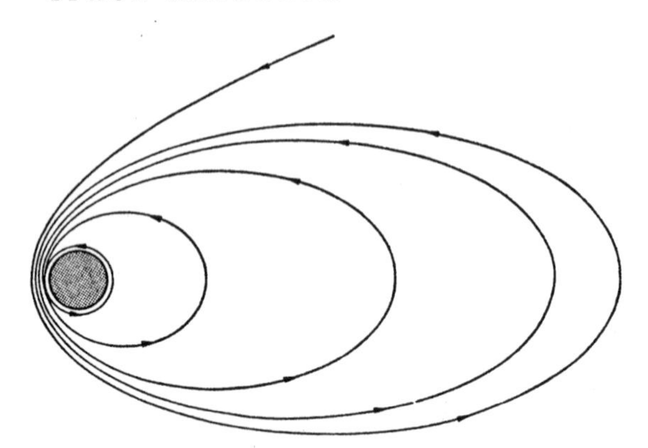

FIG. 5. Aerodynamic braking, shift from a parabolic approach orbit to a satellite orbit

Although the heating and deceleration accompanying atmospheric entry bring about severe design problems, the presence of a planetary atmosphere is advantageous in that it acts as a cushion to reduce a space vehicle's velocity to a safe landing speed. Without an atmosphere, as in the case of a landing on the Moon, one is forced to the weight-consuming expedient of rocket braking.

DECELERATION AND HEATING

A body approaching a planetary atmosphere possesses a large amount of energy; and one of the most important of the problems of atmospheric penetration is the dissipation of this energy in a manner that will not destroy the vehicle, either during penetration or on landing. If all of the vehicle's energy were converted into heat within the body itself, it would in most cases be more than sufficient to vaporize the entire body. The survival of many natural meteorites, however, is an obvious indication that not all of the energy goes into the body. Actually, the body's initial energy is transformed, through gas-dynamic drag, into thermal energy in the air around the body; and only part of this energy is transferred to the body as heat. The amount of the original energy that appears as heat in the body depends upon the characteristics of the gas flow around the body.

The chief effects accompanying atmospheric entry are reduction of vehicle velocity accompanied by appreciable deceleration and heating. Both deceleration and heating are most severe when there is a combination of high atmospheric density and high vehicle velocity, i.e., when vehicles fly at high speeds down to low altitudes. This condition is most apt to occur when the approach velocity is very high and/or when the entry is at a high angle. On the other hand, a lower initial velocity or a shallow entry angle (tangential approach) tends to restrict high velocities to higher altitudes. The initial entry velocity is determined by the planet's gravitational characteristics and by the type of vehicle mission, i.e., return from satellite orbit, return from outer space, etc.; and, therefore, one must usually just accept the initial entry velocity involved. An appropriate angle of entry, however, can be chosen to reduce the severity of entry conditions.

Deceleration can also be caused high in the atmosphere by the use of a body having high drag and/or some aerodynamic lift. High drag produces deceleration at high altitudes, while aerodynamic lift allows a more gradual descent. The slender, low-drag body shown in figure 6a would experience more severe heating and less deceleration than the blunt body in figure 6b. If, however, the latter body were pointed in the position shown in figure 6c, a lift force would be developed; and it would assume a more shallow path of descent with less heating and deceleration.

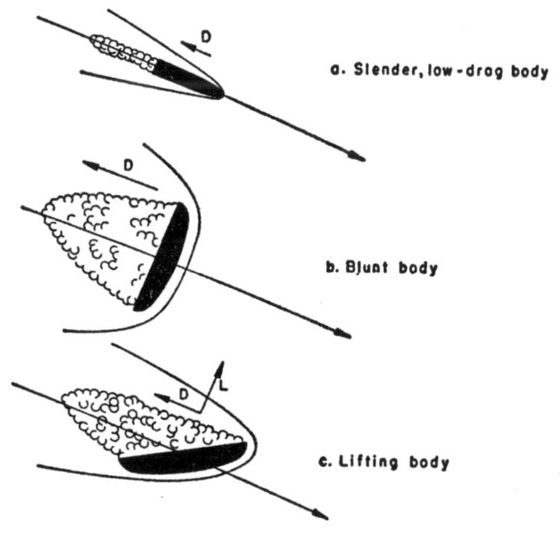

FIG. 6. Aerodynamic forces on various bodies

INFLUENCE OF PROPERTIES OF THE ATMOSPHERE

The physical and chemical characteristics of the planetary atmosphere also strongly influence entry characteristics. To gain an appreciation of the gas-dynamic forces and heating involved, a knowledge of the density variation in the atmosphere is sufficient. The approximate density variation in the Earth's atmosphere and the estimated density variation in the atmospheres of Venus and Mars are shown in figure 7. The

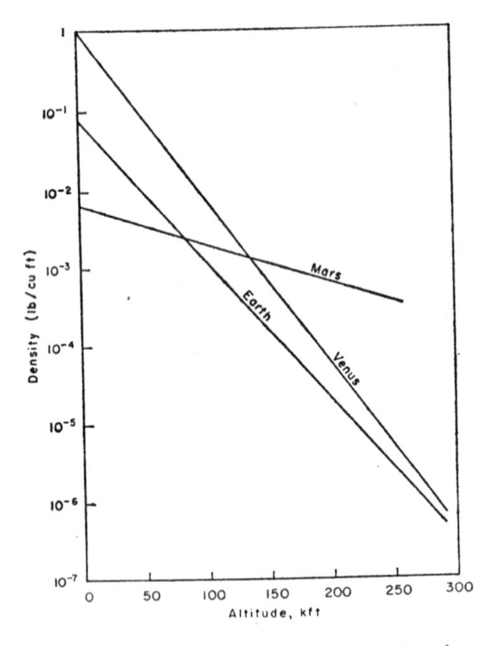

FIG. 7. Density distribution in planetary atmospheres

atmosphere of Venus, estimated to consist of about 10 percent nitrogen and 90 percent carbon dioxide, is somewhat more dense than the Earth's atmosphere, but varies in a similar way with altitude. The atmosphere of Mars, estimated to contain about 95 percent nitrogen and 5 percent carbon dioxide, is appreciably less dense than the Earth's atmosphere at surface level, but drops off much more gradually with increasing altitude and is actually more dense at high altitudes. The more gradual density variation in the Martian atmosphere effectively makes it "softer," so that it would involve a comparatively less severe entry.

A SIMPLE ANALOGY

The effects of some of these factors on atmospheric entry conditions can be visualized with the aid of a simple analogy. Imagine the problem of crash-landing a light airplane in a dense forest (figure 8). Close to the ground, the trees have thick trunks; farther up, the trunks and limbs are more slim; and at the treetops, only slender twigs and branches occur. The forest, then, is analogous to the atmosphere—dense at low altitudes and tenuous at high altitudes. When the airplane enters the trees, it suffers deceleration due to impacts with parts of the trees (aerodynamic drag) and suffers surface damage due to abrasion by twigs and branches (aerodynamic heating). If the airplane dives vertically at high speed into the forest, it will penetrate the thin upper branches without much deceleration and will still be moving at high speed when it reaches the heavy lower branches. Consequently, the deceleration and surface abrasion would be great. However, if an attempt is made to reduce speed and glide into the treetops at a low angle, the plane will decelerate more slowly in the thin upper branches and will be moving at a relatively lower speed when it finally reaches the heavy lower trunks. A still better approach could be accomplished by pulling up just before striking the treetops so that the plane's altitude tends to keep it on the tops of the trees (i. e., aerodynamic lift).

The effects of the drag characteristic of the body can be visualized by imagining the landing of two different airplanes, e.g., a modern fighter and a World War I fighter. The heavy, low-drag modern fighter would penetrate at high velocity into the heavy lower branches. The relatively light, high-drag, obsolete airplane would be decelerated with comparative comfort in the light upper branches.

The effects of the density distribution in the planetary atmosphere can be visualized by considering a different type of tree. For example, the "softer" Martian atmosphere, with a lower sea-level density and a more gradual variation of density with altitude than the Earth's atmosphere, can be visualized as a forest of taller trees with smaller trunks and a more gradual variation of branch size with height.

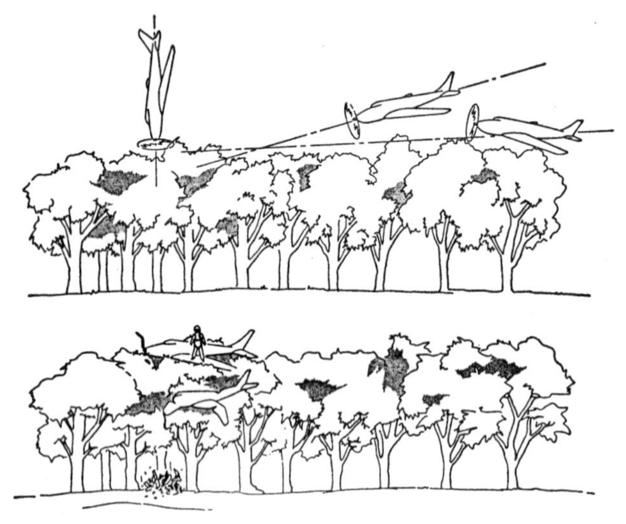

FIG. 8. A simple analogy

C. Dynamics of Atmospheric Penetration

The three general types of atmospheric penetration illustrated in figure 3—the steep descent path of a direct entry from space, the more gradual descent path of a satellite orbit decay, and a very gradual glide or lifting descent—are accompanied by differing patterns of deceleration.[9]

DIRECT DESCENT

The influence of entry angle on this deceleration pattern is shown in figure 9 for a body like the Vanguard satellite. An entry angle θ of 90° indicates a vertical descent. (θ is the angle between the vehicle path at entry and the local horizontal.) The maximum deceleration during direct entry is independent of the drag characteristics of the body. It is dependent only upon the path angle, the initial velocity, and the atmospheric characteristics. Only the altitude at which the greatest deceleration takes place is dependent upon the drag characteristics of the body.

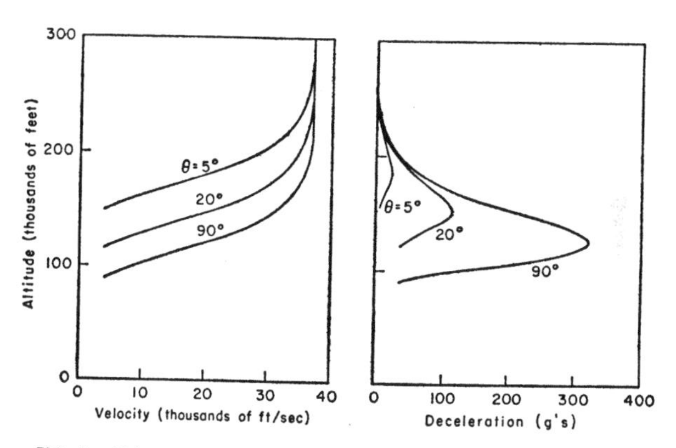

FIG. 9. Velocity and deceleration during direct entry into the Earth's atmosphere from space at various angles

$$\frac{W}{C_D A_C} = 10 \text{ lb/ sq ft}$$

The pattern of velocity and deceleration for the same body is shown in figure 10 for vertical penetration of the atmospheres of Venus, Earth, and Mars. The different gravitational attractions account for the different initial velocities, which are equal to the *escape velocities*. The effect of atmospheric density variation is apparent in the shape and position of the curves. The more dense atmosphere of Venus results in deceleration at a higher altitude; however, the velocity variation with altitude and the maximum deceleration are about the same as for Earth because of the similar atmospheric density pattern. The more gradual variation of density in the Martian atmosphere results in a more gradual variation of velocity with altitude and a lower peak deceleration.

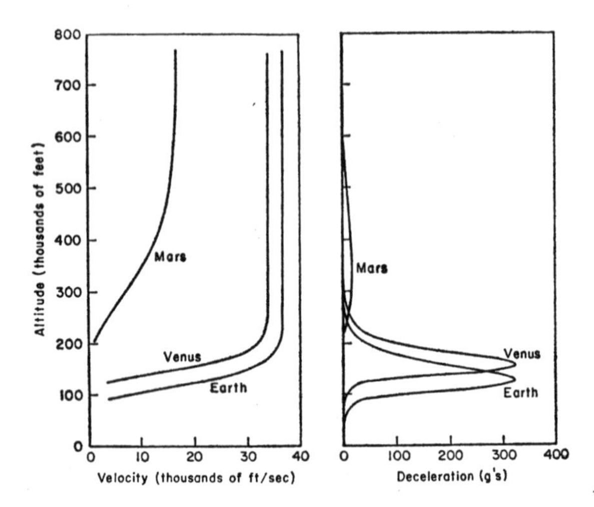

FIG. 10. Velocity and deceleration during direct entry from space at $\theta = 90°$ into three planetary atmospheres

$$\frac{W}{C_D A_C} = 10 \text{ lb/ sq ft}$$

The phase of satellite orbital decay of interest here is the portion where heating and deceleration are greatest—say, about the last 2,000 miles and the last few minutes of the satellite's lifetime. This phase is preceded by a much longer period covering many revolutions in which the satellite executes a very gradual spiral that becomes more circular. The rate of energy loss by the vehicle due to aerodynamic drag is small enough so that the vehicle's kinetic energy (energy of motion) and potential energy (energy of height) adjust themselves to a momentary "equilibrium" orbit. In this process, the potential energy decreases, and the kinetic energy increases. Thus the satellite's velocity actually increases in the beginning phases of orbital decay.

The deceleration pattern in the last phase of descent is very similar to that for direct descent at a very shallow angle. For the Vanguard example used above, the peak deceleration would be about 9 g.

LIFTING DESCENT

A lifting descent involves a still more gradual atmospheric penetration, and here again the path angle adjusts itself to the forces acting on the vehicle and is generally quite small—something like a few tenths of a degree. In this case, the deceleration does not go through a sharp peak, but increases gradually to a maximum. Decelerations can be limited to rather small values in a lifting descent.

RANGE AND TIME OF DESCENT

It should be noted that the more gradual descents involve longer times and cover greater ranges than the steeper descents. For example, starting at the same altitudes and velocities, a direct descent may cover a distance of only a few hundred miles and be accomplished in about one-half minute; an orbital decay might cover a range of a few thousand miles in 5 or 10 minutes; and a lifting descent might extend over 5,000 to 10,000 miles in about 2 hours. A gradual descent involves reduction of velocity and consequent loss of energy over a long period of time.

EXAMPLES OF DECELERATION LOADS

Some examples of deceleration loads that would be experienced in various kinds of entries are listed in table 1. These are to be compared with an allowable load tolerance of roughly 10 to 15 g for manned vehicles.

HEATING DURING ATMOSPHERIC PENETRATION

The reduction of the vehicle's kinetic and potential energy during descent is accompanied by heating of the surrounding air; some of this heat is transferred to the vehicle's surface. The fraction of this kinetic and potential energy which reaches the vehicle surface as heat is of primary concern to the designer. This fraction, or *conversion efficiency,* depends upon the vehicle's shape and upon its velocity and altitude—and ultimately upon the manner of heat transfer between the hot gas and the vehicle surface. At very high altitudes, the heat

TABLE 1.—*Maximum deceleration experienced during various types of atmospheric penetration*

[Values given in Earth g's]

Planet	Direct entry at escape velocity			Direct entry at orbital velocity			Entry by decay from satellite orbit	Entry of lifting vehicle at orbital velocity		
	$\theta = 5$	20	90	$\theta = 5$	20	90		L/D = 1	2	5
Venus	28.6	112	326	14.3	56	163	8.9	0.88	0.44	0.18
Earth	28.3	111	324	14.2	55.5	162	9.5	1.0	.5	.2
Mars	1.6	6.3	18.3	.8	3.2	9.2	1.4	.38	.2	.07

energy is developed directly at the vehicle's surface, and one-half the vehicle's lost energy appears as heat in the body. At lower altitudes, thermal energy appears in the air between the shock wave and the body. Heat is transferred from this hot air to the body by conduction and convection through a viscous *boundary layer*. Radiation from the hot gas may also contribute appreciably to the surface heating.

D. Temperature, Heating Rates, and Payload Protection

The heating of a vehicle in a given application determines the type of surface-protection system it needs.[10, 11]

Supersonic aircraft and hypersonic glide missiles operate with essentially constant skin (surface) temperatures, and the vehicle must be so designed that heat is carried off at the same average rate as it is acquired. The temperature attained by various parts of the body will depend on the radiative characteristics of the vehicle surface and the local heating rates.

During the flight of a ballistic missile, the skin temperature experiences a large variation. During ascent through the atmosphere, the skin experiences moderate heating—about like that for a supersonic aircraft. Heating drops to zero upon exit from the atmosphere, and the skin is cooled by radiating its heat while the missile is going over the top of its trajectory. During descent, heating increases enormously and the skin reaches its highest temperature sometime during this re-entry part of its flight.

SURFACE COOLING SYSTEMS

Methods for protecting a payload from high external heating include—

Thickening of skin, to absorb heat in a greater mass of material in the case of sudden heating and to compensate for decreased material strength at elevated temperatures in the case of steady-state or equilibrium conditions.

Insulation of the outer surface, to reduce transmission of external heat into the payload compartment and structure. Also, the higher surface temperatures reduce surface heating because of increased radiative cooling from the surface.

Cooling of the inner skin surface, absorbing transmitted heat by bailing internal water or heating a coolant fluid.

Transpiration cooling, pumping of gas or vapor through a porous skin to carry heat away from and insulate the vehicle.

Ablation cooling, carrying away heat and insulating by vaporizing the surface material of the vehicle shell.

Combinations of these.

The choice of the most effective method of protection against re-entry heating for a given mission must of course rest upon detailed design studies.

VISUAL PHENOMENA

Some idea of the spectacular nature of a high-speed re-entry can be gained from the following excerpt from an account of the re-entry of the nose cone and associated structures from the firing of an Army Jupiter missile on May 18, 1958: [12]

> Comdr. R. G. Brown, captain of the U.S.S. *Stickell,* was the first to spot the re-entry phenomena. At the time he spotted the light, it appeared about as bright as a star of third magnitude. The appearance occurred almost exactly where the corrected position was predicted.
>
> Within 3 seconds after the first re-entry light was observed the phenomena had blossomed into 3 distinct objects. The brightest object appeared similar to a huge magnesium flare, which was assumed to be the booster. The light emitted by this object definitely pulsated as if the body were tumbling through space. At one time, the body's trajectory was nearly in line with the planet Jupiter. It was estimated that the brightness was at least 1,000 times that of the planet.
>
> The second brightest visual object was a beautiful blue-green. This was assumed to be the instrument compartment. The blue-green light may have resulted from the copper and magnesium in this section. The actual trajectory of the instrument compartment during its burning stage was much shorter than the booster and nose cone. Slightly past the midpoint of the visible trace, the blue-green light turned nearly white and then burned out.
>
> The nose cone never reached a white color. The radiation in the visible spectrum was orange-red in color. Visibility of the nose cone, spacewise, began slightly behind the booster, and then moved ahead of the booster. During the last few seconds of the visible flight, the booster and nose cone moved behind

the large cumulus cloud to the south of the U.S.S. *Stickell.* The radiation was so intense that the whole cloud became illuminated. It was in this section of the flight that the booster ceased to glow and became invisible. The nose cone was seen to appear from behind the cloud and was tracked for another couple of seconds before it cooled enough to become invisible. The total time of visibility from the position of the U.S.S. *Stickell* was approximately 27 seconds.

Notes

[1] Williams, E. P., and Carl Gazley, Jr., Aerodynamics for Space Flight, The RAND Corp., Paper P-1256, February 24, 1958.

[2] Williams, E. P., et al., Long-Range Surface-to-Surface Rocket and Ramjet Missiles—Aerodynamics, The RAND Corp., Rept R-181, May 1, 1950.

[3] See footnote 2.

[4] See footnote 2.

[5] Sänger, E., and I. Bredt, Uber einen Raketenantrieb für Fernbomber, ZWB, UM Nr. 3533, Berlin, 1944. (Available as Navy translation CGD-32, A Rocket Drive for Long-Range Bombers.)

[6] Goldberg, P. A., Electrical Properties of Hypersonic Shock Waves and Their Effect on Aircraft Radio and Radar, Boeing Airplane Co., Report No. D2-1997, July 2, 1957.

[7] Sisco, W. B., and J. M. Fiskin, Effect of Relatively Strong Fields on the Propagation of EM Waves Through a Hypersonically Produced Plasma, Douglas Aircraft Co., Report No. LB-25642, November 22, 1957.

[8] Roberts, C. A., W. B. Sisco, and J. M. Fiskin, Theory of Equilibrium Electron and Particles Densities Behind Normal and Oblique Shock Waves in Air, Douglas Aircraft Co., Report No. LB-25872, September 1, 1958.

[9] Gazley, C., Jr., Deceleration and Heating of a Body Entering a Planetary Atmosphere from Space, Vistas in Astronautics, Pergamon Press, 1958 (proceedings of the First Annual Air Force Office of Scientific Research Astronautics Symposium).

[10] Gazley, C., Jr., Heat-Transfer Aspects of the Atmospheric Re-entry of Long-Range Ballistic Missiles, The RAND Corp., Rept. R-273, August 1, 1954.

[11] Masson, D. J., and Carl Gazley, Jr., Surface-Protection and Cooling Systems for High-Speed Flight, Aeronautical Engineering Review, vol. 15, No. 11, November 1956.

[12] Woodridge, D. D., and R. V. Hembree, Operation Gaslight, Jupiter Missile AM-5, Army Ballistic Missile Agency, Huntsville, Ala., June 5, 1958.

14

Landing and Recovery

A. Determining Factors

Landing by manned space vehicles is subject to the same considerations that apply to landings by aircraft, parachutes, etc. Thus this discussion is largely devoted to landings by unmanned vehicles.

Landing a payload on the Earth, Moon, or distant planetary body or asteroid is influenced mainly by[1]

Velocity of approach to the landing surface.
Nature of "target" material.
Nature of the payload.
Weight limitations.

B. Velocity of Approach

The *velocity of approach* is the velocity of the payload just prior to contact with the "target" material. The kinetic (motion) energy possessed by the payload by virtue of this velocity must be completely dissipated in landing. The kinetic energy is a fundamental basis for comparison of impacts—a heavy body striking at low velocity is about equivalent to a light body striking at high speed, if their kinetic energies are equal.

The physics of impact suggests that the dissipation of energy can take place in many ways—and, in fact, the precise redistribution of energy cannot be predicted with current theory for even the simplest impact problem. Heat, friction, deformation of both target and payload, wave energy imparted to the surrounding atmosphere (if any) are some of the physical phenomena that take place in most impacts.

C. Nature of Target

Landings on the Earth may take place in water or on land. Landings may occur on a variety of surfaces, such as dry sand, clay, or rocks of varying degrees of consolidation. Problems of landing in crushed rock, shale, or pebbly material, with each piece relatively hard but free to move, are very different from those associated with landing on an outcropping of solid rock.

Terrains are classified as to their compressive strength, determined by firing projectiles into material samples, and by compression tests in the laboratory. These are rather crude approximations to the very wide spectrum of terrain conditions on Earth and, most probably, on other planetary bodies. However, a considerable uncertainty in compressive strength is tolerable for properly designed landing systems.

D. Nature of Payload

The nature of the payload determines the degree of complication required in the landing operation.

Experiments indicate that amplifiers, transistors, and other electronic gear can be designed and mounted to withstand very heavy deceleration loads—3,000 to 20,000 g or more. These are well above loads encountered in the majority of controlled landings. Consequently, excluding exceptionally delicate instruments, design specifications for landings are not likely to be dictated by the fragility of electronic equipment.

One extreme limit on landing speed is provided by the need to keep communication antennas above ground—the payload should not hit so hard that it buries itself completely.

The landing loads tolerable by human beings are among the lowest likely to be specified in the design of landing systems.

E. Weight Limitations

There is a considerable weight difference between equipment for landing on planets having atmospheres and equipment for landing on bodies such as the Moon that have no appreciable atmosphere. Presence of an atmosphere permits use of lifting surfaces and other aerodynamic devices like parachutes for gradual descent to a light landing. Without an atmosphere, braking rockets, which are extravagant in their use of available weight capacity, are required to reduce the approach velocity to a permissible level.

F. Controlled Landing in Soils

Typical designs for landing on soils are depicted in figures 1 to 3. They all aim to reduce the landing load by taking up the shock in extended penetration of the target surface by a body of relatively small cross-section.

For terrain that is not level, a direct impact perpendicular to the ground surface cannot, of course, be guaranteed.

Spherical payloads with several spikes attached to a vehicle (figure 2) have been suggested to increase the chance that in any orientation one spike will imbed. Figure 1 illustrates the notion of using one or more devices for several purposes—if practicable. Thus, for a lunar landing, the *retrorocket* case can deform upon impact and partially dissipate kinetic energy. The soft landing vehicle of figure 3 is more complex and bulky but has the advantage that the spider legs, which may incorporate energy-dissipating devices, would help position a landing vehicle appropriately for special purposes, such as takeoff for a return trip (figure 4).[2]

Landings are classified, somewhat arbitrarily, as soft or hard, depending upon whether approach velocities are less or greater than 500 feet per second, respectively.

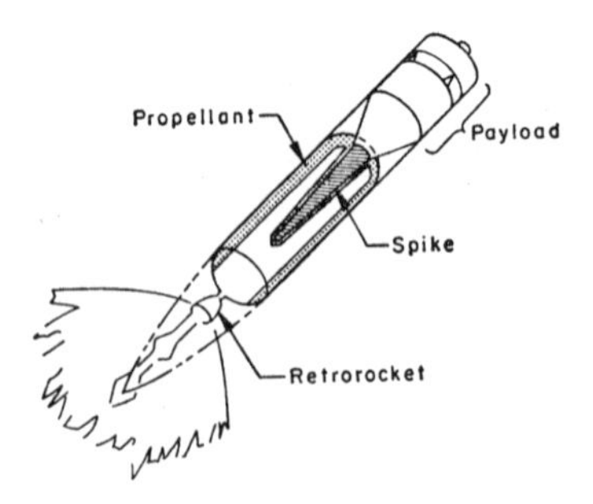

FIG. 1. Impact vehicle with penetration spike

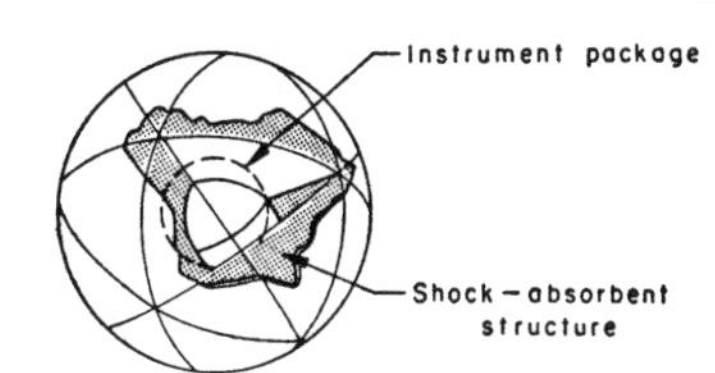

FIG. 2. Rough-landing instrument carrier
(spikes not shown)

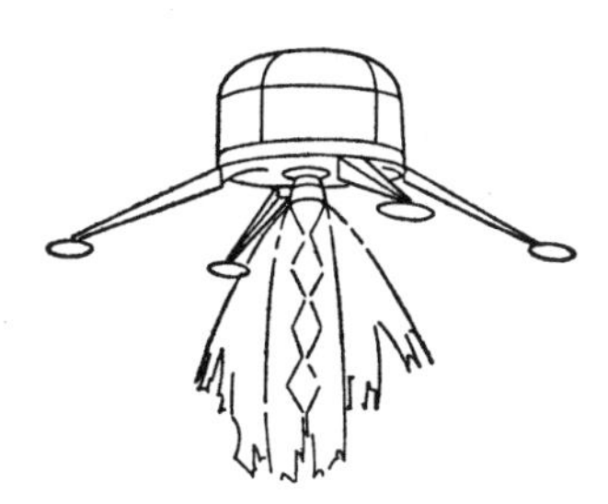

FIG. 3. Soft-landing vehicle

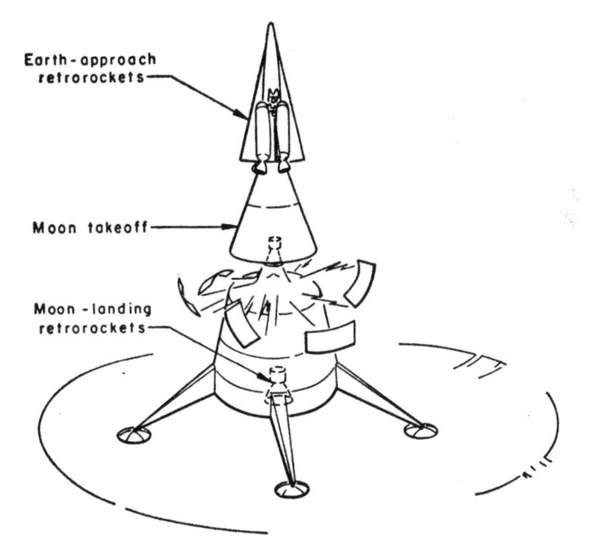

FIG. 4. Soft-landing vehicle, full assembly

G. Water Landing

No planetary body other than Earth is known to be covered to any appreciable extent by fluids. Consequently, at present, landing in water is of importance only in recoveries on the Earth.

Water impacts may occur either because it is difficult to control the landing operation to strike within a designated land area or because weight limitations preclude the use of control devices.

Where a controlled landing is possible, it may be desirable to land in water, because there is a high probability of recovering the payload from such a landing. Impacts on land involve the hazards of burial of payload and difficulty of access and search in some areas because of climate, terrain, or political factors.

Some other advantages of water recovery are these:

> The behavior of objects striking water can be easily tested experimentally, and the target material is not variable like terrain.
>
> The uprightness of the payload after landing is easily guaranteed by packaging it to be heavy on the "down" end.

Water recovery, of course, requires the payload to be waterproof, and the payload or some auxiliary attachment to be floatable.

Notes

[1] Lang, H. A., Lunar Instrument Carrier—Landing Factors, The RAND Corp., Research Memorandum RM-1725, June 4, 1956.

[2] Figures 1 to 4 are adapted from G. A. Olson, Lunar Vehicles, Proceedings of Lunar and Planetary Exploration Colloquium, July 15, 1958, pp. 10-11.

15

Environment of Manned Systems

A. Internal Environment of Manned Space Vehicles

Available information on the important internal environmental aspects of manned space vehicles and their effect on human occupants is gathered here from a large number of sources in the literatures of aviation medicine, submarine and deep-sea-diving medicine, and human heat regulation.[1-34]

To function properly, any system must be maintained within certain physical and chemical limits. Man is no exception.

The primary elements of the environment bearing on the health and well-being of the man are the following:

Composition and pressure of the atmosphere.
Gravitational forces (acceleration).
Temperature.
Radiation.

Of course, there are other factors, such as noise and vibration during takeoff and landing phases, possible hazards from foreign toxic substances, nutritional requirements, effects of confinement and isolation, and the like, which must be examined separately.

In general, if one graphically shows the variation of a limiting stress or departure from some preferred environment with time, a curve results which separates the graph into two general areas, as in figure 1. The area below the curve represents conditions under which man can operate more or less normally and efficiently. The area above the curve represents conditions which the man cannot tolerate and under which he cannot operate. Because of the severity of the stress in this area, he is either incapable of performing, or is unconscious, or is dead. Separating these two regions is a broad transition band representing (for any given individual) a gradual loss of efficiency and loss of ability to recover promptly when the stress is removed. The deeper we penetrate into the transition band, the greater the physiological strain.

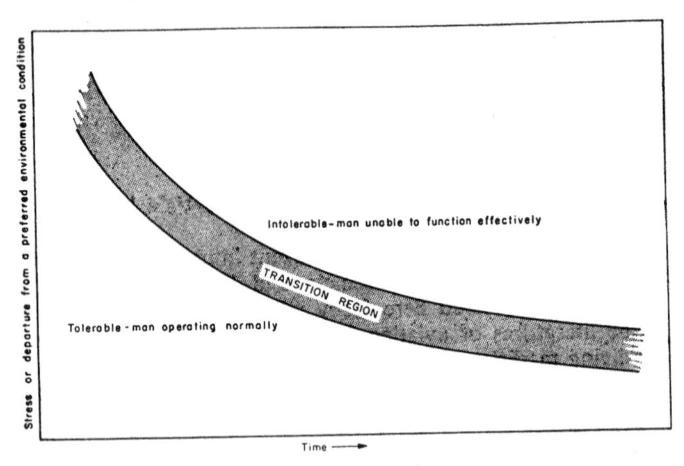

FIG. 1. Stress-time effects on humans

The following general observations are significant:

Individuals differ greatly in their ability to withstand stress. For some individuals the critical region would be moved down and to the left. For some individuals it would be moved up and to the right.

A given individual will not always react in the same way at different times. In addition, he can improve his performance by gradual acclimatization or conditioning.

The curves can be modified by all of the other factors in the man's environment.

And, of course, the milder the stress the longer he can tolerate it.

COMPOSITION AND PRESSURE OF THE ATMOSPHERE

First, the effect of the gaseous environment and respiratory requirements (figure 2). This figure represents the region of man's tolerance to variations in the partial pressure of oxygen in the air he breathes. The normal sea-level value is 149 millimeters of mercury. The lower band in the figure represents the minimum oxygen partial pressure that can be tolerated in the air entering the lungs. Man's reaction depends, of course, both upon the pressure of the oxygen he breathes and the rate of penetration into the intolerable region.

For gradual penetration into the region of too little oxygen, the usual symptoms are sleepiness, headache, lassitude, altered respiration, psychologic impairment, inability to perform even simple tasks, and eventual loss of consciousness.

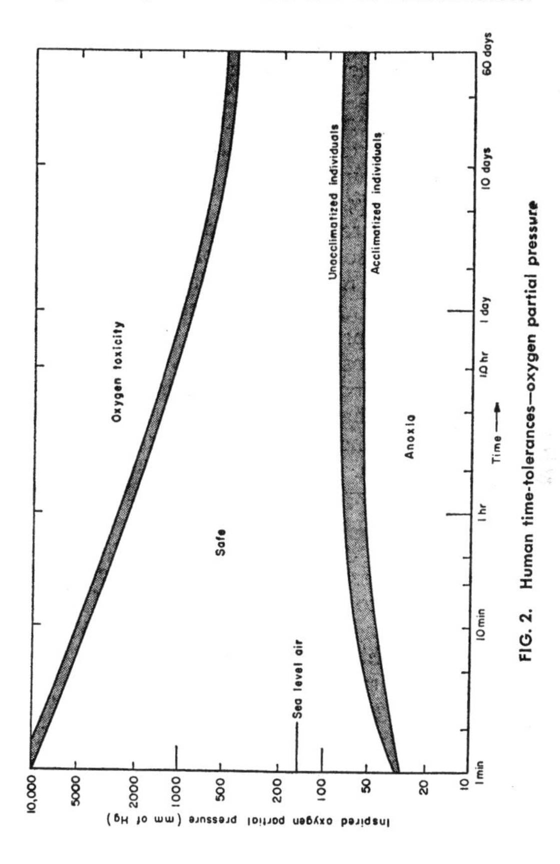

FIG. 2. Human time-tolerances—oxygen partial pressure

For a sudden transition into excessively low oxygen pressures, the intermediate symptoms are bypassed and the man rapidly loses consciousness, goes into spasms or convulsions. If the low oxygen pressure is accompanied by low total pressure, the sudden decompression is characterized by pains in the chest and in the joints due to bubble formation (or bends) and by confusion, delirium, and collapse.

Too much oxygen can also be lethal, as indicated by the upper curve, which bounds the oxygen toxicity region, the symptoms again depending upon rate of onset. Prolonged exposure to one atmosphere of pure oxygen, for example, eventually produces inflammation of the lungs, respiratory disturbances (coughing, gasping, and pulmonary congestion), various heart symptoms, numbness of the fingers and toes, and nausea.

Exposure to still higher partial pressures of oxygen produces nervousness, discomfort, irritation of the eyes and virtual blindness, nausea, loss of consciousness, and convulsions. In general, for long exposures, inspired (inhaled) oxygen partial pressures should be kept well within the extremes of 80 and 425 millimeters of mercury to avoid undesirable effects. It should be mentioned also that man's tolerance to other stresses (e. g., acceleration) is reduced by a low oxygen pressure.

It is perhaps easier to appreciate the respiratory requirements of men if they are expressed in terms of the total pressure and atmospheric composition (figure 3). Here total pressure in pounds per square inch is plotted against the volume percentage of oxygen in the surrounding "air." Normal sea-level conditions are 14.7 pounds per square inch absolute (p. s. i. a.) and 21 percent oxygen. Proceeding from this point, we may reach the critical low-oxygen region by several different routes. For example, if we maintain the total pressure at 14.7 p. s. i. a. and reduce the concentration of oxygen, a critical point for the unacclimatized person is reached when the oxygen concentration has fallen to about 11 percent. Or, if we keep the composition of the air constant at 21 percent oxygen and reduce the pressure (which is what happens when we ascend to higher altitudes in our atmosphere), a critical point is reached when the surrounding pressure has fallen to about 8.4 p. s. i. a. This corresponds to air at about 15,000 feet. An equivalent condition for a man breathing an atmosphere of pure oxygen would be reached when the ambient pressure was about 2.5

p. s. i. a., which corresponds to the ambient pressure at about 42,000 feet.

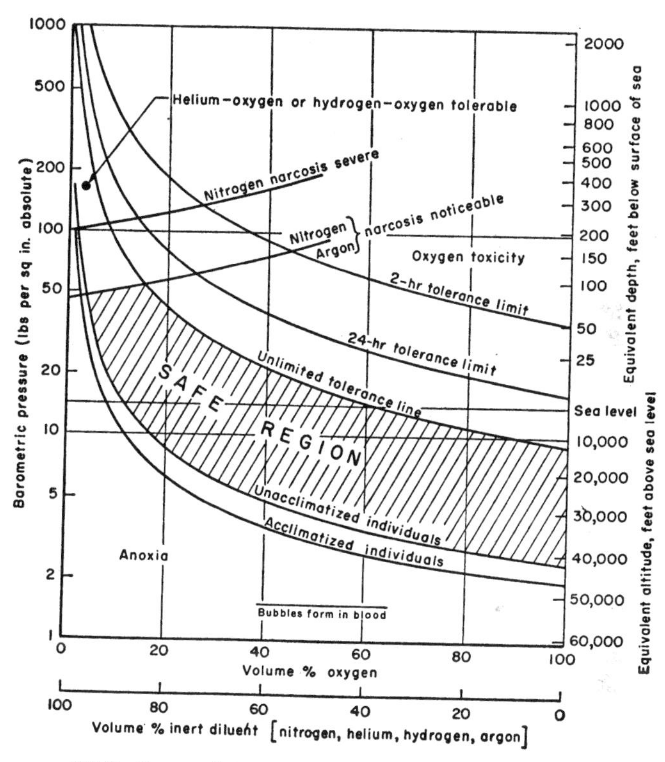

FIG. 3. Human tolerances—atmospheric composition and pressure

Harmful effects of the inert ingredient (diluent) do not become apparent until the total pressure reaches several atmospheres (40-50 p. s. i. a.). If the inert diluent is nitrogen, above this pressure nitrogen narcosis is encountered, the effects of which have been compared to alcohol intoxication: confusion, disorientation, loss of judgment, and unconsciousness. However, if helium is the inert diluent, much higher total pressures may be tolerated without any adverse effects, up to possibly 20 atmospheres total pressure.

Since man normally produces carbon dioxide at nearly the same rate as he consumes oxygen, this is an important environmental consideration (figure 4).

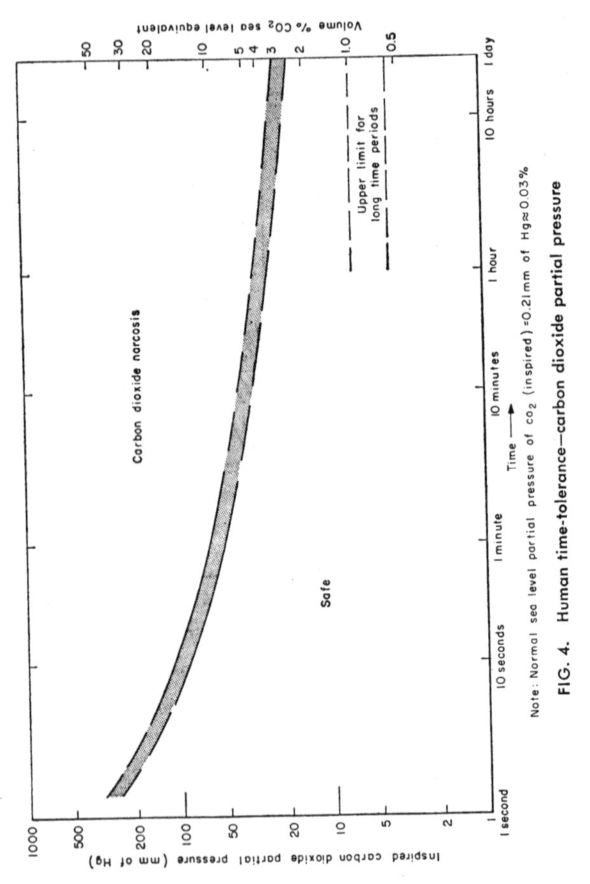

Note: Normal sea level partial pressure of CO_2 (inspired) = 0.21 mm of Hg ≈ 0.03%

FIG. 4. **Human time-tolerance—carbon dioxide partial pressure**

Above 15 to 20 millimeters of mercury partial pressure of carbon dioxide in the inspired air (equivalent to 2 to 3 percent carbon dioxide at 1 atmosphere), the intermediate effects are a noticeable increase in the breathing rate, and distention of the air sacs of the lungs, with impairment of the normal gas exchange in the lungs. Partial pressures above about 35 millimeters of mercury (equivalent to 5 percent carbon dioxide at 1 atmosphere) can be tolerated for a period of only a few minutes. Acute symptoms are heavy panting, marked respiratory distress, fatigue, stupefaction, narcotic effects, unconsciousness, and eventual death. For exposure for long periods of time most authorities recommend that the inspired carbon dioxide pressure be kept below 4 to 7 millimeters of mercury (0.5 to 1 percent at 1 atmosphere pressure).

GRAVITATIONAL FORCES

The relative *position* or orientation of the subject is of first importance in determining tolerable levels of gravitational or acceleration force, or "g force." As the g force is gradually increased, certain effects are observed (table 1).

TABLE 1.—*Gross effects of acceleration forces*

Effects:	*g's*
Weightlessness	0
Earth normal (32.2 feet/second ^{2})	1
Hands and feet heavy; walking and climbing difficult	2
Walking and climbing impossible; crawling difficult; soft tissues sag	3
Movement only with great effort; crawling almost impossible	4
Only slight movements of arms and head possible	5

Positive longitudinal g's, short duration (blood forced from head toward feet):

Effects:	*g's*
Visual symptoms appear	2.5-7.0
Blackout	3.5-8.0
Confusion, loss of consciousness	4.0-8.5
Structural damage, especially to spine	>18-23

Transverse g's, short duration (head and heart at same hydrostatic level):

Effects:	*g's*
No visual symptoms or loss of consciousness	0-17
Tolerated	28-30
Structural damage may occur	>30-45

Figure 5 shows the time-tolerance relationships for positive longitudinal (lengthwise) and transverse (crosswise) forces experienced while the subject is either prone or supine, prone being the position of lying face down and supine being the position of lying face up.

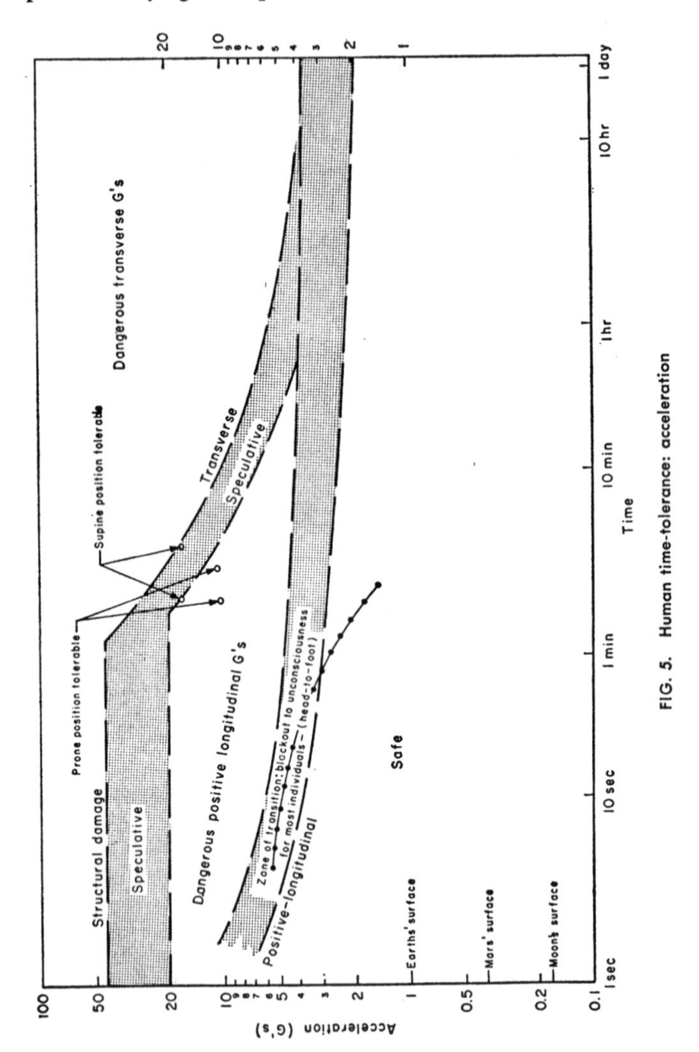

FIG. 5. Human time-tolerance: acceleration

For the transverse position, human subjects in Germany during World War II are reported to have been subjected to 17 g's for as long as 4 minutes with no harmful effects and no loss of consciousness. The curves indicated for very long periods of time are best estimates and thus speculative, since no data are available on long-term effects. Col. John Stapp, Air Force Missile Development Center, has investigated extreme g loadings, up to 45 g's, sustained for fractions of a second. These are the kinds of accelerations or decelerations that would be experienced in crash landings. For these brief high g loadings, the rate of change of g is important, injuries becoming more probable when rate of change of g exceeds 500 g's per second.

As a matter of interest, the beaded line on the figure indicates the approximate accelerations that would be experienced by a man in a vehicle designed to reach escape velocity with three stages of chemical burning, each stage having a similar load-factor-time pattern. This curve enters the critical region for positive g's. Most individuals would probably black out and some would become unconscious. However, for individuals in the transverse position, this acceleration could be tolerated and the individual would not lose consciousness.

It is important to note that there is no compelling reason for adopting elaborate measures either in equipment design or in flight crew selection to achieve extraordinarily high g force tolerances. Virtually all rocket engines are capable of being throttled to reduce thrust, and thereby vehicle acceleration; in this way g forces can be limited to any reasonable value (say, 4 or 5 g).

High g forces are more likely to be an important factor in the re-entry of "capsules" with no lifting surfaces. Glide vehicles can limit re-entry loads to very modest levels.

WEIGHTLESSNESS

Only a very limited body of information is available concerning the effects of complete absence of g force. The longest periods of *weightlessness* so far experienced by human beings have been something like 40 to 50 seconds in aircraft on zero g flight trajectories. As with other forms of stress, different people react in different ways to weightlessness. Some individuals find it unpleasant and some seem to enjoy it, but the durations so far have all been short. At present it is believed that at least some individuals will be able to adapt to a weightless condition for long periods of time. In any event, there are ways of avoiding weightlessness through vehicle rotation, if it is found to be a generally undesirable condition.

The longest space flight to date by an animal was that of the dog in Sputnik II. The vehicle was under zero g, but the acceleration force actually experienced by the animal was dependent upon the rotation rate of the satellite and the location of the dog relative to the axis of rotation—and these factors are not known with certainty at present.

TEMPERATURE

Temperature is a more familiar variable element in man's environment—also a more complicated one to depict in simple terms (figure 6). In the high-temperature region of the chart, for example, relative humidity is as important as temperature in its effect on human tolerance. It is well known that much higher temperatures can be tolerated if the humidity is low. Other important factors are thermal radiation, wind velocity, acclimatization, and the amount of activity being engaged in. The curves presented here assume that adequate drinking water is being supplied to make up for water lost in perspiration, which can amount to very sizable quantities at high temperatures.

The general symptoms of excessively high temperature, depending upon degree and length of exposure, are lassitude, loss of efficiency, weakness, headaches, inability to concentrate, apathy, increase in cardiac work, nausea, visual disturbances, increased oxygen uptake, increased body temperature, heat stroke, and convulsions.

In the low-temperature region, the situation is also highly complicated and the position of the critical region is affected by the amount of activity, thermal radiation, wind velocity, acclimatization, and the insulating effectiveness of gloves and boots. The curves of figure 6 suggest a limitation for a man wearing the warmest clothing that he can wear and still retain the ability to move freely (clothing about 1 inch thick). The usual hazards from low temperature are cold injury to extremities, frostbite, exhaustion, local freezing, depression of respiratory and circulatory functions, general hypothermia, paralysis, and death.

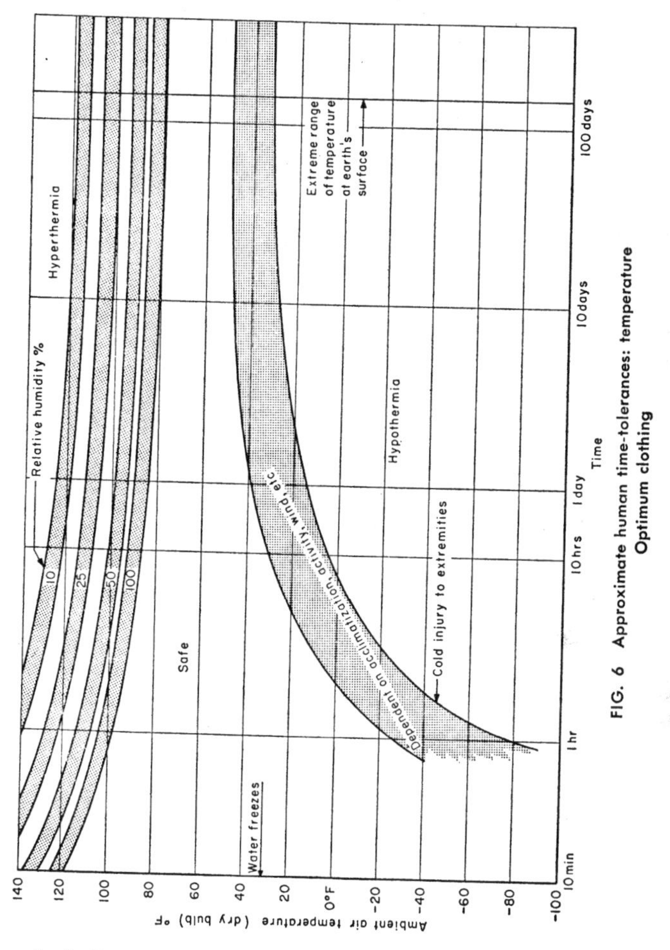

FIG. 6 Approximate human time-tolerances: temperature
Optimum clothing

As indicated above, temperature has a profound effect on the quantity of water required per man per day. It has a less marked effect on the food and oxygen requirements, as indicated in figure 7. These curves are intended to be roughly suggestive of the dependence of water, food, and oxygen consumption upon surrounding air temperature. They are greatly affected by the amount of work activity, and also by such factors as the amount of clothing worn, humidity, acclimatization, and body weight. However, as indicated, the quantity of water required becomes of overriding importance at temperatures higher than about 70° F. Thus, in a space ship it would be desirable to maintain the living quarters at a comfortable temperature, not only for the comfort of the crew but also to avoid excessive loading of the water-removal or water-recycling equipment.

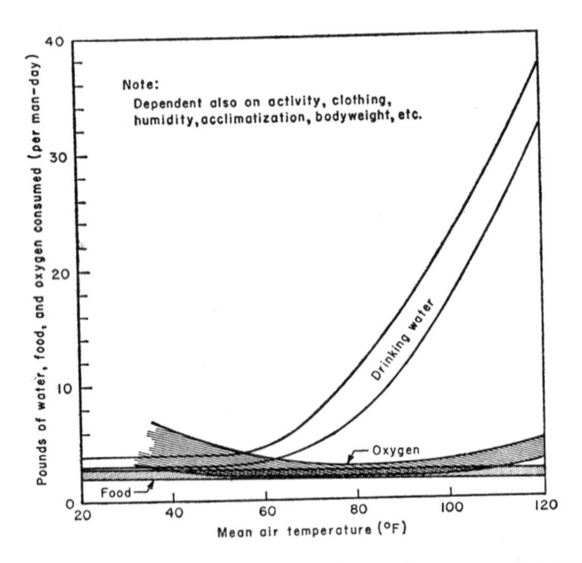

FIG. 7. Effect of temperature on water, food, and oxygen requirements

With respect to the problem of recovering man's metabolic waste products for reuse in a sealed cabin, many possible schemes have been proposed but none has yet been perfected. The actual type of scheme that would be used would depend upon the length of the projected excursion into space.

RADIATION

First it should be pointed out that the forms of radiation from the Sun other than the occasional radiation from solar flares can be adequately handled within today's technology. The effects of thermal radiation can be controlled by adjusting the *absorptivity* and *emissivity* of the outer skin of the vehicle, and almost any desired skin temperatures can be obtained. As for solar radiation in the visible, ultraviolet, and soft X-ray regions, present data indicate that these do not constitute a direct hazard to crews of space vehicles, as they can easily be stopped or weakened by thin layers of almost any structural material.

The newly discovered "radiation belts" of the Earth present a problem that can be met either by avoiding them or by shielding to reduce dosages to human beings to acceptable levels. This radiation is apparently X-rays produced when high-velocity charged particles strike the material of which a space vehicle is constructed (figure 8); it is known as *bremsstrahlung,* or "impact radiation." Manned satellites, to avoid the radiation belts, could orbit the Earth at altitudes lower than three or four hundred miles. The occupants of space vehicles escaping from the Earth could be shielded with fairly thin sheets of dense materials such as lead; or escape routes over the Earth's polar regions might be used to avoid the radiation belts almost entirely. More information about peak dose rates in these belts is needed to establish the best procedures for dealing with the problem, but solutions are available.

However, the hazard of cosmic radiation remains an open question, since there is no satisfactory way of shielding against it. What is still unknown is the *relative biological effectiveness* (RBE) of cosmic radiation as compared with other forms of radiation we know more about.

Figure 9 displays the generally accepted human tolerances with respect to genetic and nongenetic damage by gamma radiation. For example, for persons in the reproductive ages, the recommended dosage limits for genetic reasons are not more than 300 milliroentgens per week, no more than 15 roentgens in any one year, and no more than a 5-roentgen-per-year average dose. A man can, however, tolerate much larger dosages without any noticeable nongenetic effects (up to 50 roentgens in an acute dose).

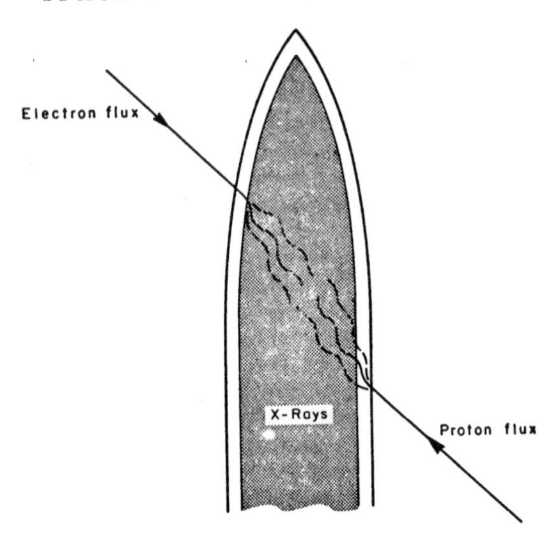

FIG. 8. Production of x-rays inside space vehicle due to Bremsstrahlung caused by protons and electrons

What would be the effects on humans of the heavy nuclei of cosmic radiation? It all depends upon the RBE, and at present the data are inadequate to calculate a meaningful RBE. Physical measurement indicates that the dosage that would be received from unshielded cosmic radiation is low.

The commonly accepted view is that there will be no measurable adverse effects for short exposures but possibly some genetic effects or minor local tissue effects when the exposure is prolonged.

So far, in short-duration experiments with rats, there has been no evidence of cancer produced by cosmic radiation, although black mice and guinea pigs exposed to cosmic radiation at high altitudes for 24 to 36 hours subsequently developed gray spots in their hair, indicating permanent damage to parts of hair follicles by cosmic heavy nuclei.

The whole question of the effects of cosmic radiation is now being studied intensively and undoubtedly much will be learned in the next few years.

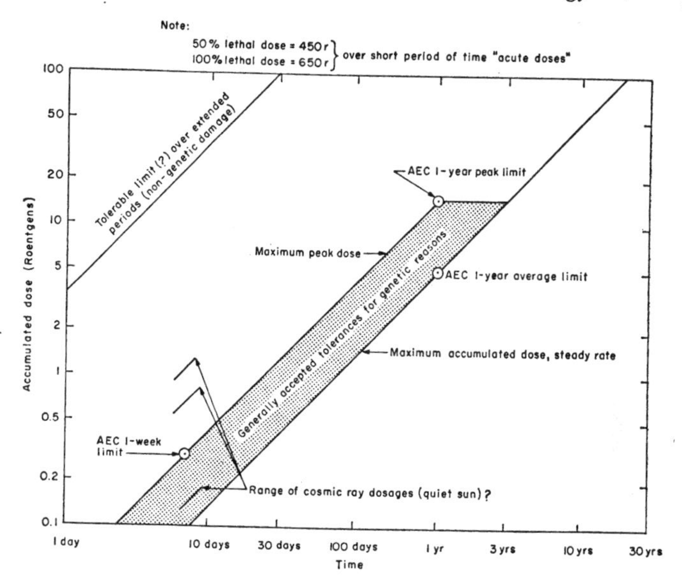

FIG. 9. Human tolerances to radiation

AIR SUPPLY

A 5-man crew will require approximately 100 pounds of oxygen for a 10-day trip.[35, 36] Three possibilities immediately suggest themselves: liquid oxygen (LOX),[37] hydrogen peroxide, and some form of plant life to regenerate the air.

Liquid oxygen normally requires a double-walled insulated storage container, approaching in weight the LOX itself. Also water must be added before the resulting gaseous oxygen can be breathed.

Hydrogen peroxide (H_2O_2) does not present so difficult a storage problem, and is somewhat less hazardous. Using 90 percent H_2O_2, 236 pounds (plus a 40-pound container) would be required to yield 100 pounds of oxygen. The rest of the weight (136 pounds) would appear as water. This would itself be useful, and consequently cannot be counted against a hydrogen peroxide system. Furthermore, energy released during the decomposition of H_2O_2 to oxygen and water could be used as an auxiliary power source.

For very long flights one would attempt to establish a close regenerative cycle similar to that which exists in nature—man consuming oxygen and producing carbon dioxide, while plants reverse the cycle. Algae appear attractive because they have a high photosynthetic efficiency, have no waste stalks, etc.[38] However, algae tanks are usually cumbersome and may require considerable power (about 1,400 watts per man).

The inert component of a space vehicle atmosphere is most likely to be nitrogen.[39] Suitable means must be provided for filtering and purifying the atmosphere and controlling the moisture content.

B. Food Preservation

Aboard space craft on extended trips it will be desirable for the morale and health of the crew to provide food that is varied in form and of high quality. Water may be recycled, and possibly food also, through the use of algae or various synthetic processes. Nevertheless, homegrown food would undoubtedly be desirable. Satisfactory methods of preserving food should (1) not make them distasteful to humans; (2) insure a long storage life, preferably without refrigeration; (3) not require bulky or heavy packaging. Concerning the last, it may also be desirable to lighten the weight of the product further by dehydration.

In addition to conventional canning, freezing, and pickling[40] there are three new approaches to food preservation: (1) gamma irradiation, (2) beta irradiation, and (3) freeze-drying. Referring to the first two, gamma rays or electrons are used to extend the storage life of foods by inhibiting sprouting and destroying microorganisms, parasites, or insects. Food preserved by irradiation is subjected to very little temperature rise —normally enzymes are not deactivated unless a very strong dose is administered.

Radiation treatment has been under intensive investigation at the University of Michigan Fission Products Laboratory, and at the Quartermaster Food and Container Institute in Chicago. The latter also subcontracts work. Sixty-nine cooperative Army research-contract holders are listed in the footnote reference below.[41] Gamma irradiation can be provided from a number of sources: spent reactor fuel elements, a reactor core surrounded by a blanket containing a liquid with a specific gamma producer, separated fission products, gaseous fission products, and artificial isotopes. As long as there are no neutrons mixed with the radiation, and the energy of the latter is below the neutron *binding energy* in the target, there should be no induced radioactivity.[42]

Typical doses required for specific applications are as follows:[43] onion, potato, and carrot sprout-inhibition, 5,000 to 15,000 roentgen-equivalent-physical [44] (r. e. p.); mold inhibition on citrus, 150,000 to 250,000 r. e. p.; trichina irradiation, 50,000 r. e. p.; insect deinfestation of grain, 50,000 to 100,000 r. e. p.; pasteurization, 500,000 to one million r. e. p.; sterilization, 2 to 4 million r. e. p. Potatoes given 7,000 r. e. p. have resisted sprouting for up to 5½ months at room temperature,[45] and well over a year when refrigerated.[46, 47] Raw ground pork treated with rather high dosages has a refrigerated storage life of 10 to 11 days.[48] Irradiated apples may be kept in a refrigerator for several months (although the radiation tends to lower the total pectin content).

Beta radiation may be supplied by *Van de Graff machines, linear accelerators,* etc. Electrons do not penetrate very well, but this may be an advantage in certain cases—e.g., surface mold treatment of peaches or citrus fruit. However, the Army is constructing at Stockton, Calif., a 20-m. e. v. (million electron volts) linear accelerator capable of treating slabs of food up to 6 inches thick (both sides are irradiated simultaneously). Induced radioactivity may be a problem at these energies.

In the case of freeze drying, food is first frozen, then placed in a vacuum and subjected to a pulsed electromagnetic beam (of radar frequencies) to sublime the ice crystals (that is, to cause the ice to go directly to the vapor state without passing through the liquid or water state). (If a continuous beam were used, the center might be cooked.) The resulting product will have lost approximately 90 percent of its weight, and both bacterial and enzyme actions are inhibited by the absence of moisture. Refrigeration is not necessary if air- and moisture-proof packaging is available. The Raytheon Co. reports having successfully applied this technique to mushrooms, carrots, beef ribs, steak, veal cutlets, pork chops, lobster, shrimp, fish, strawberries, and peas. In the case of shrimp, the product has the consistency of popcorn. Preparation of the dehydrated shrimp for eating requires one-half hour of soaking in tepid water, and 2 minutes in boiling water. Although not entirely overlapping, it is clear that this process is in a sense competitive with the other two.

Now let us compare some of the relative advantages and disadvantages of canning, freezing, freeze drying, and irradiation.

CANNING

Advantages: Generally no refrigeration required, and a long storage life.

Disadvantages: Low acceptability, increased weight, and high container cost.

FREEZING

Advantages: High acceptability, medium packaging costs, extended storage life as long as freezing temperatures are maintained.

Disadvantages: Freezers are bulky, expensive, and to some extent unreliable.

FREEZE DRYING

Advantages: No refrigeration required (if moisture-vapor and oxygen can be sufficiently excluded by the package), long storage life, light product weight, possibly high acceptance. Vitamins and the protein structure remain intact.

Disadvantages: Problems of packaging. Even as much as 2 percent moisture by weight will cause "browning"—a nonenzymic chemical reaction. A slightly greater moisture content will activate enzymes and then bacteria.

IRRADIATION

Advantages: Versatile, can increase the refrigerated storage life of meat and produce, as well as break the trichinosis cycle. If sterilization were possible, refrigeration might be avoided altogether.

Disadvantages: Only a relatively few items can be treated by irradiation without producing undesirable tastes, colors or odors, and generally the stronger the dosage the worse the effects. Sterilizing doses almost always produce undesirable side effects—broccoli, for example, turns gray and limp.

C. Choice of Propellants for Launching

An important distinction may exist between liquid and solid propellant rockets for launching manned vehicles. Experience to date with liquid rockets shows that virtually all destructive failures ("blowups") are accompanied by rather slow buildup of fires with wide dispersal of the propellants. A manned vehicle built to survive re-entry conditions could probably be expected to sustain the havoc created by such launching rocket failures. On the other hand, bursting with explosive suddenness is not an uncommon type of malfunction with solid rockets, and survival chances may not be good in the event of such a launching rocket failure. Future tests of all sorts of rockets should be observed for evidence of their relative suitability for launching manned vehicles.[49]

D. Artificial Gravity

As a general rule, machinery can be fairly readily designed for operation in a weightless environment. The effects of this environment on humans, however, is not known, although brief tests made in aircraft are encouraging.

It is often suggested that, should weightlessness prove to be a problem, it can be solved by "artificial gravity" caused by rotating the space vehicle. This is possible since rotation produces centrifugal acceleration that can be made equal, at any desired point, to the acceleration of normal gravity. There are, however, certain limitations that must influence vehicle design and may also lead to a need for training and adaptation.[50] First of all, centrifugal acceleration, for a fixed rate of rotation, depends on the distance from the axis of rotation and always acts outward from that axis. If a man were to lie across the center of a rotating space vehicle with his head on one side of the axis of rotation and his feet on the other, he would be simply stretched by the centrifugal force—weightlessness might be more agreeable. For rotation to provide a reasonable substitute for gravity the design must be such that the man is at a distance from the axis of rotation that is large compared with his length; i.e., he must be several multiples of 6 feet away from the axis, say, 100 feet. Even at a point 100 feet from the rotation axis, the effective g force varies about 6 percent over the length of a 6-foot man, his feet experiencing a greater force than his head.

Another point of consideration arises from the fact that a man in a space vehicle will want to move, and a moving body will experience some odd effects in a rotating vehicle due to the *Coriolis force*. This force exists because the vehicle is a rotating reference frame for the man's movements and, therefore, differs from the customary nonrotating environment of everyday experience. As a result of Coriolis effects, a man will experience a level of g force that depends upon his direction of movement. If he is 100 feet from the rotation axis and the average g force is the normal sea-level value, he will experience a g force variation of almost 20 percent between a slow walk in one direction and a slow walk in the opposite direction. If he walks toward or away from the axis of rotation, he will feel a g force of about one-tenth normal in a sidewise direction. These effects suggest that a man in the artificial gravity of a rotating vehicle may stagger a bit until he gets his "space legs."

Notes

[1] Adolph, E. F., and Associates, Physiology of Man in the Desert, Interscience Publishers, Inc., New York, 1947.

[2] Armstrong, H. G., Principles and Practice of Aviation Medicine, 3d ed., The Williams & Wilkins Co., Baltimore, 1952.

[3] Berry, C. A., The Environment of Space in Human Flight, Institute of the Aeronautical Sciences, New York, Preprint No. 796, January 1958.

[4] Buettner, K., Man and His Thermal Environment, Mechanical Engineering, November 1957.

[5] Burnight, T. R., Physics and Medicine of the Upper Atmosphere, ch. 13, Ultraviolet Radiation and X-rays of Solar Origin, University of New Mexico Press, Albuquerque, 1952.

[6] Burton, A. C., and O. G. Edholm, Man in a Cold Environment, Physiological and Pathological Effect of Exposure to Low Temperatures, Edward Arnold (Publishers), Ltd., London, 1955.

[7] Carlson, L. D., Man in Cold Environment, a Study in Physiol-

ogy, Alaskan Air Command, Ladd Air Force Base, Fairbanks, Alaska, August 1954.

[8] Code, C. F., et al., The Limiting Effect of Centripetal Acceleration on Man's Ability To Move, Journal of Aeronautical Science, vol. 14, 1947, p. 117.

[9] Cranmore, Doris, Behavior, Mortality, and Gross Pathology of Rats under Accelerative Stress, Aviation Medicine, April 1956.

[10] Dill, D. B., Life, Heat, and Altitude, Physiological Effects of Hot Climates and Great Heights, Harvard University Press, Cambridge, Mass., 1938.

[11] Flight Surgeons Manual, Air Force Manual 160-5, Department of the Air Force, U. S. Government Printing Office, Washington, D. C., 1954.

[12] Gagge, A. P., Man's Response to Temperature Extremes, Proceedings of the Fifth AGARD General Assembly, Ag 20/P10, June 1955, pp. 82-91.

[13] Gagge, A. P., et al., The Influence of Clothing on the Physiological Reactions of the Human Body to Varying Environmental Temperatures, American Journal of Physiology, vol. 124, 1938, pp. 31-50.

[14] German Aviation Medicine, World War II, vols. I and II, prepared under the auspices of the Surgeon General, USAF, Department of the Air Force, U. S. Government Printing Office, Washington, D. C., 1950.

[15] Haldane, J. B. S., What Is Life? Boni and Gaer, New York, 1947.

[16] Harrington, L. P., A Survey Report on Human Factors in Undersea Warfare, ch. 13, Temperature and Humidity in Relation to the Thermal Interchange between the Human Body and the Environment, Committee on Undersea Warfare, National Research Council, Washington, D. C., 1949.

[17] Krieger, F. J., A Casebook on Soviet Astronautics, pt. II, The RAND Corp., Research Memorandum RM-1922, June 21, 1957.

[18] Lee, D. H. K., Human Climatology and Tropical Settlement, University of Queensland, Brisbane, 1947.

[19] Lewis, C., USAF School Simulates Living in Space, Aviation Week, January 27, 1958, pp. 49-61.

[20] Ley, Willy, and W. von Braun, The Exploration of Mars, The Viking Press, New York, 1956.

[21] Mayo, A. M., Survival Aspects of Space Travel, Aviation Medicine, October 1957.

[22] McFarland, Human Factors in Air Transportation, McGraw-Hill Book Co., New York, 1956.

[23] Simons, David G., and Druey P. Parks, Climatization of Animal Capsules during Upper Stratosphere Balloon Flights, Jet Propulsion, July 1956, pp. 565-568.

[24] Stapp, J. P., Collected Papers on Aviation Medicine, AGARD-

ograph No. 6, ch. 14, Tolerance to Abrupt Deceleration, Butterworth Scientific Publications, London, 1955.

[25] Stauffer, F. R., Acceleration Problems of Naval Training—1: Normal Variations in Tolerance to Positive Radial Acceleration, Aviation Medicine, June, 1953, p. 167.

[26] Stewart, W. K., Some Observations on the Effect of Centrifugal Force in Man, Journal of Neurological Psychiatry, vol. 8, 1945, p. 24.

[27] Stewart, W. K., Lectures on the Scientific Basis of Medicine, II, The Physiological Effects of Gravity, London, 1952-53, pp. 334-343.

[28] Stoll, A. M., Human Tolerance to Positive G as Determined by the Physiological End Point, Aviation Medicine, August 1956, p. 356.

[29] Submarine Medicine Practice, Bureau of Medicine and Surgery, Department of the Navy, NAVMED-P 5054, U. S. Government Printing Office, Washington, D. C., 1956.

[30] Webster, A. P., Acceleration Limits of the Human Body, Aviation Age, March 1956, p. 26.

[31] Zuidema, G. D., et al., Human Tolerance to Prolonged Acceleration, Wright Air Development Center, WADC TR 56-406, October 1956.

[32] Nadel, A. B., Human Factors Requirements of a Manned Space Vehicle, General Electric Technical Military Planning Operation, Rept. No. RM 58TMP-10, April 10, 1958.

[33] Air University Quarterly, summer issue, 1958.

[34] Hullinghorst, Col. R. L., Meeting the Physiological Challenge of Man in Space, Army Information Digest, vol. 13, No. 10, October 1958, p. 38.

[35] Ross, H. E., Orbit Bases, Journal of the British Interplanetary Society, vol. 8, No. 1, January 1949, pp. 1-19.

[36] Paris, N. S., and S. S. Naistat, Hydrogen Peroxide as a Source of Oxygen, Water, Heat, and Power for Space Travel, Proceedings of the American Astronautical Society, Fourth Annual Meeting, January 29-31, 1958, New York, pp. 31-1 to 31-13.

[37] Compressed air itself would not be particularly desirable. Man uses up oxygen from the air, replacing it with carbon dioxide. Therefore it is sensible to take along only that constituent which is being consumed.

[38] Gaume, J. G., Design of an Algae Culture Chamber Adaptable to a Space Ship Cabin, Proceedings of the American Astronautical Society, Fourth Annual Meeting, January 29-31, 1958, New York, pp. 22-1 to 22-4.

[39] Haviland, R. P., Air for the Space Ship, General Electric Co., Document No. 56SD235; reprinted from Journal of Astronautics, vol. 3, No. 2, summer 1956.

[40] In canning, bacteria are killed and enzymes deactivated by

heat; in pickling, the high pH concentration prevents bacteria growth; and in freezing, the low temperature inhibits bacteria growth.

[41] The Interdepartmental Radiation Preservation of Food Program, February 15, 1957, the Interdepartmental Committee on Radiation Preservation of Food.

[42] Selection of a Food Irradiation Reactor Type—Phase 1, Internuclear Co., Inc., Report AECU-3319, July 1, 1956.

[43] Morgan, B. H., G. E. Donald, G. E. Tripp, and D. F. Farkas, Basic Concepts in the Application of Ionizing Radiations to Foods for Preservation, Paper No. 57-NESC-117, Second Nuclear Engineering and Scientific Conference, March 11-14, 1957, ASME.

[44] Defined as 93 ergs absorbed per gram of tissue.

[45] Bialos, I., Potatoes May Be First Food Preserved by Atomic Energy, Western Grower and Shipper, June 1955.

[46] There are some anomalies in sprout inhibition. A dose of radiation sufficient to inhibit sprouting by Spanish onions may accelerate it in the case of white pearl onions.

[47] Brownell, L. E., and S. N. Purohit, Combining Gamma Radiation, Refrigeration, Refrigeration Engineering, June 1956.

[48] See footnote 43.

[49] Goldsmith, M., Suitability of Solid and Liquid Rocket Engines for Placing Manned Satellites in Orbit, The RAND Corp., Paper P-1542, November 10, 1958.

[50] Lawden, D. E., The Simulation of Gravity, Journal of the British Interplanetary Society, vol. 16, No. 3, July-September 1957, p. 134.

16

Space Stations and Extraterrestrial Bases

A. Space Stations

The importance of a large, permanent, manned Earth satellite was recognized quite early in the history of rocket flight. In 1929, H. Oberth[1] considered many of the observational functions of such a satellite. At about the same time, a series of papers by Von Pirquet [2] stressed the significance of an orbital assembly station for space-flight missions. Since that time numerous articles have been published on the use, construction, and operation of space stations and specialized space vehicles. These range in quality from pure science fiction to detailed engineering computations. At least one article meriting consideration appeared in Science Wonder Stories.[3] Some general descriptive material may be found in engineering literature.[4-13]

Many of the functions that could be performed by a space station could also be performed by Earth satellites which are neither large, permanent, nor manned. The unique function of the space station is to serve as a base for large space vehicles. The Earth is not an entirely satisfactory base, due in part to the presence of the atmosphere, which presents a serious re-entry problem, and in greater part to the strong gravitational field at the surface of the Earth.

Present rockets, and the larger ones that will become possible when development is completed on announced engines of 1.5-million-pound thrust, can send very substantial payloads into space trajectories. However, these will still fall far short of the needs of more advanced space missions. One partial solution is to develop ever larger chemical boosters or large boosters of higher performance using nuclear propulsion. Nevertheless, the net capability of the mission will always be limited by the thrust of the largest booster system available at any given time. Space missions such as manned expeditions to the neighboring planets would have to await many generations of missile development. The space station offers an alternative solution.

This method of operation would involve several steps. Individual shipments of fuel, guidance equipment, structural material, parts, tools, personnel, and supplies would be transported from the Earth to the space station, with the size of each shipment limited by the performance of available boosters. The accumulation of components may proceed as long as necessary, and the feasible total is, therefore, essentially unlimited. Long-range spacecraft would be assembled on, and depart from, the station. For certain missions the spacecraft may be returned to the station and used repeatedly. A considerable potential advantage of the space station lies in the fact that space vehicles departing from it would have less severe structural and propulsion restrictions than rockets that must depart from and return to the Earth.

Various proposals for designing and constructing a space station have been advanced in the past.[14-24] The more realistic studies show that a minimum payload on orbit is necessary before a station may be constructed. Beyond this threshold a station of virtually any size may be assembled.

It is relevant to note that ICBM-type boosters with added stages appear to be about adequate for the construction of a space station. Large booster systems that can be based on 1.5-million-pound engines are certainly adequate. There is not yet sufficient information for a decision as to whether or not a station should be built.

For staging of interplanetary flights, the orbit of the space station would lie in the plane of the Earth's orbit around the Sun. For long life, its altitude would be above 300 miles. To avoid dangerous radiation from the "radiation belt" it might be below 500 miles or possibly above 30,000 miles. Considerations such as these tend to limit the usefulness of the station for some secondary functions which might be performed better by less elaborate satellites, manned or unmanned.

In addition to being a staging base for equipment, the space station would also be a transfer point for personnel. Crews assigned to long space voyages could undergo a preliminary period of adaptation to the space environment aboard the station before departure. The station may also be used for quarantine. Returning spacecraft and personnel may possibly carry diseases or other forms of contamination, and should not be permitted to return directly to Earth before careful observation. These functions seem characteristic of a space station used as a staging base, and may be regarded as a representative, rather than exhaustive, list.

Two outstanding problems to be overcome in establishing a space station are the difficulty of rendezvous between the station and supply rockets, and the propulsion requirement that individual shipments be of sufficient size to permit construction to begin. Additional problems include the details of fabrication, the provision for adequate power supply, and consideration of such factors affecting personnel as air, food, water, radiation, weightlessness, and collision with foreign particles.

The minimum practical value for individual supply loads delivered into orbit seems to be about 3,000 to 4,000 pounds.

This would allow carriage of a man or men, sufficient supplies to last until resupply, communication and guidance equipment to assist in rendezvous, and some rocket propulsion to alter the orbit. At least one package of tools and construction material may be placed in orbit first and the manned vehicle sent later. Although this minimum value may be theoretically adequate, it appears quite marginal with very little provision for emergencies. In particular, this weight is not likely to include provision for a safe return to earth if rendezvous fails to occur before the supplies are exhausted. On the other hand, an individual package weight of about 10,000 pounds could sustain a small crew for a considerable time and include adequate safety provisions. Something approaching an orbiting payload of 10,000 pounds is achievable with an ICBM booster augmented by a fairly large upper stage.[25]

The rendezvous problem is in a more uncertain state, although informed opinion suggests that it can be solved by known techniques. Propulsion and controls must be provided to reduce position and velocity differences between the two vehicles to a sufficiently small value to permit mechanical contact by a beam or cable.

It appears, in sum, that the construction of a space station of some sort will be technologically feasible in the foreseeable future.

B. Extraterrestrial Bases

Bases, more or less permanent in character, will be required at some point in an expanding space program. A good deal of attention has already been given to the matter of a base on the moon.[26-37] Bases of one kind or another will also be required if manned interplanetary travel is to develop in a serious way—first of all because it would be grossly inefficient to remain, say, 1 day on Mars, after taking half a year to get there; and, second, because return from a flight to, again, Mars, may be really feasible only after a delay of many months to bring the Earth and Mars into favorable relative positions for the return flight.

For human beings to exist in a base on the Moon or one of the planets, the internal environment must be generally the same as that in a space station. One important difference in the two cases, however, is that the local environment of the

Moon or planet must be coped with, and that local assets may be exploited. Generally speaking, the establishment and maintenance of a manned base on any extraterrestrial body would be a very difficult and ambitious undertaking. The problems associated with a base on the Moon are less difficult than they would be for bases on any other body, since the Moon is relatively close to the Earth at all times. Moreover, the average temperature of the Moon (notwithstanding the tremendous day-night variations) is not far from the average temperature of the Earth. Thus, in a well-insulated base on the Moon, temperature regulation would not constitute a severe problem. If efficient conversion of solar energy is assumed and if usable local sources of water and oxygen can be obtained through chemical processing of native materials, a partially self-sufficient Moon base can be envisioned.

Second in order of difficulty would come Mars as a site for a manned base. More remote than the Moon but providing a more nearly earthlike environment in many ways, Mars has a thin atmosphere which would provide protection against meteorites and some slight protection against abrupt temperature changes. Present knowledge of the surface barometric pressure is inadequate to say whether a human being outside a base would require a full-pressure suit or whether a partial-pressure suit would be adequate. In any event, a breathable atmosphere would have to be provided, and a manned base would have to be airtight with artificial heating supplied throughout most of the daily cycle.

None of the other planetary bodies of the solar system, in the light of present knowledge, offers an attractive site for an extraterrestrial base. Venus, although often cited as a possibility, contains such high concentrations of carbon dioxide in its atmosphere and has such a dense cloud cover that establishment of a base on its surface appears exceedingly more difficult than on Mars. More information about Venus is needed, however, before definitive statements about its suitability for a base can be made.

It has been suggested that extraterrestrial residences may be a long-term solution to the problem of accommodating the enormously increasing population of the earth. This notion, while not entirely out of the question, does exert something of a strain on credibility. However, it can probably be fairly asserted that developments for extraterrestrial bases may help by disclosing ways of dealing with unfavorable environments and thereby effectively enlarging the amount of usable land area on the Earth.

Notes

[1] Oberth, Hermann, Wege zur Raumschiffahrt (Methods of Achieving Space Flight), 3d ed., R. Oldenbourg, Berlin, 1929.

[2] Von Pirquet, Guido, Die Rakete (The Rocket), Journal of the Verein für Raumschiffahrt, Breslau, 1927-29.

[3] Noordung, Hermann, The Problems of Space Flying, revised English printing in Science Wonder Stories, 1929.

[4] The Station in Space, Journal of the American Rocket Society, vol. 63, September 1945, pp. 8-9.

[5] Clark, A. C., Interplanetary Flight, Temple Press, London, 1950, especially ch. VIII.

[6] Fears, F. D., Interplanetary Bases—The Moon and the Orbital Space Station, Journal of Space Flight, vol. 3, September 1951, pp. 4-5.

[7] Firsoff, V. A., Artificial Satellites Explained, Flight, vol. 60, October 1951, pp. 504-506.

[8] Ketchum, H. B., A Preliminary Survey of the Constructional Features of Space Stations, Journal of Space Flight, vol. 4, October 1952, pp. 1-4.

[9] Von Braun Offers Plan for Station in Space, Aviation Age, vol. 18, 1952, pp. 61-63.

[10] Von Braun, W., The Early Steps in the Realization of the Space Station, Journal of the British Interplanetary Society, vol. 12, January 1953, pp. 23-24.

[11] Ehricke, K. A., Engineering Problems of Manned Space Flight, Interavia, vol. 10, July 1955, pp. 506-511.

[12] Hoover, G. W., Sectional Satellites, Missiles and Rockets, vol. 2, October 1957, pp. 135-137.

[13] Ehricke, K. A., Our Philosophy of Space Missions, Aero Space Engineering, vol. 17, May 1958, pp. 38-43.

[14] Ross, H. E., Orbital Bases, Journal of the British Interplanetary Society, vol. 8, January 1949, pp. 1-19.

[15] Engel, R., Earth Satellite Vehicles, Interavia, vol. 5, 1950, pp. 500-502.

[16] Gatland, K. W., A. M. Kunesch, and A. E. Nixon, Fabrication of the Orbital Vehicle, Journal of the British Interplanetary Society, vol. 12, November 1953, pp. 274-285.

[17] Dixon, A. E., K. W. Gatland, and A. M. Kunesch, Fabrication of the Orbital Vehicle, in Space-Flight Problems, Laubscher & Co., Zurich, Switzerland, 1953, pp. 125-135.

[18] Ehricke, K. A., A New Supply System for Satellite Orbits, pt. I, Jet Propulsion, vol. 24, September-October 1954, pp. 302-309.

[19] Romick, D. C., Preliminary Engineering Study of a Satellite

Station Concept Affording Immediate Service with Simultaneous Steady Evolution and Growth, presented at 25th annual meeting of the American Rocket Society, November 14-18, 1955, preprint No. 274-55.

[20] Romick, D. C., Concept for Meteor—A Manned Earth-Satellite Terminal Evolving from Earth to Orbit Ferry Rockets, in Proceedings of the VII International Astronautical Congress, September 1956, pp. 335-380.

[21] Romick, D. C., R. E. Knight, and S. Black, Meteor Jr.: A Preliminary Design Investigation of a Minimum Sized Ferry Rocket Vehicle of the Meteor Concept, in Proceedings of the VIII International Astronautics Congress, 1957, pp. 340-372.

[22] Goodyear Proposes Smaller Space Station, Aviation Week, vol. 67, October 1957, pp. 115-119, 123.

[23] Clark, E., Convair Plans Four-Man Space Station, Aviation Week, April 1958, pp. 26-28.

[24] Astronautics and Space Exploration, hearing before the Select Committee on Astronautics and Space Exploration, 85th Cong., 2d sess., on H. R. 11881, April 15 through May 12, 1958; K. A. Ehricke, p. 632.

[25] See footnote 24.

[26] Comments with Respect to the Glenn L. Martin Study on Requirements for a Manned Station on the Moon, Outer Space Propulsion by Nuclear Energy, hearings before the subcommittees of the Joint Committee on Atomic Energy, Congress of the United States, 85th Cong., 2d sess., 1958.

[27] Thompson, G. V. E., The Lunar Base, Journal of the British Interplanetary Society, vol. 10, No. 49, March 1951.

[28] Fears, F., Interplanetary Bases, Journal of Space Flight, vol. 3, September 1951.

[29] Moore, P., Guide to the Moon, W. W. Norton & Co., Inc., New York, 1953.

[30] Ryan, C. (editor), Conquest of the Moon, The Viking Press, New York, 1953.

[31] Ryan, C. (editor), Conquest of Mars, The Viking Press, New York, 1953.

[32] Von Braun, W., The Mars Project, University of Illinois Press, Urbana, 1953.

[33] Clarke, A. C., Exploration of the Moon, Frederick Muller, Ltd., London, 1954.

[34] Sowerby, P. L., Structural Problems of the Lunar Base, Journal of the British Interplanetary Society, vol. 13, No. 36, January 1954.

[35] Awdry, G. E. V., Developments of a Lunar Base, Journal of the British Interplanetary Society, vol. 13, No. 16, May 1954.

[36] Sholto-Douglas, J. W. E. H., Farming on the Moon, Journal of the British Interplanetary Society, vol. 15, No. 17, January 1956.

[37] Holbrook, R. D., Outline of a Study of Extraterrestrial Base Design, The RAND Corp. Research Memorandum RM-2161, April 22, 1958.

17

Nuclear Weapon Effects in Space

A. Nuclear Weapon Effects on Personnel

In addition to the natural radiation dangers which will confront the space traveler, we must also consider manmade perils which may exist during time of war. In particular, the use of nuclear weapons may pose a serious problem to manned military space operations. The singular emergence of man as the most vulnerable component of a space-weapon system becomes dramatically apparent when nuclear weapon effects in space are contrasted with the effects which occur within the Earth's atmosphere.

When a nuclear weapon is detonated close to the Earth's surface the density of the air is sufficient to attenuate (weaken) nuclear radiation (neutrons and gamma rays) to such a degree that the effects of these radiations are generally less important than the effects of blast and thermal (heat) radiation. The relative magnitudes of blast, thermal, and nuclear radiation effects are shown in figure 1 for a nominal fission weapon (20 kilotons) which is exploded at sea level.[1]

The solid portions of the three curves correspond to significant levels of blast, thermal, and nuclear radiation intensities. Blast *overpressures* of the order of 4 to 10 pounds per square inch will destroy most structures. Thermal intensities of the order of 4 to 10 calories per square centimeter will produce severe burns to exposed persons. Nuclear radiation dosages in the range of 500 to 5,000 roentgens are required to produce death or quick incapacitation in humans.

If a nuclear weapon is exploded in a vacuum—i.e., in space—the complexion of weapon effects changes drastically:

First, in the absence of an atmosphere, blast disappears completely.

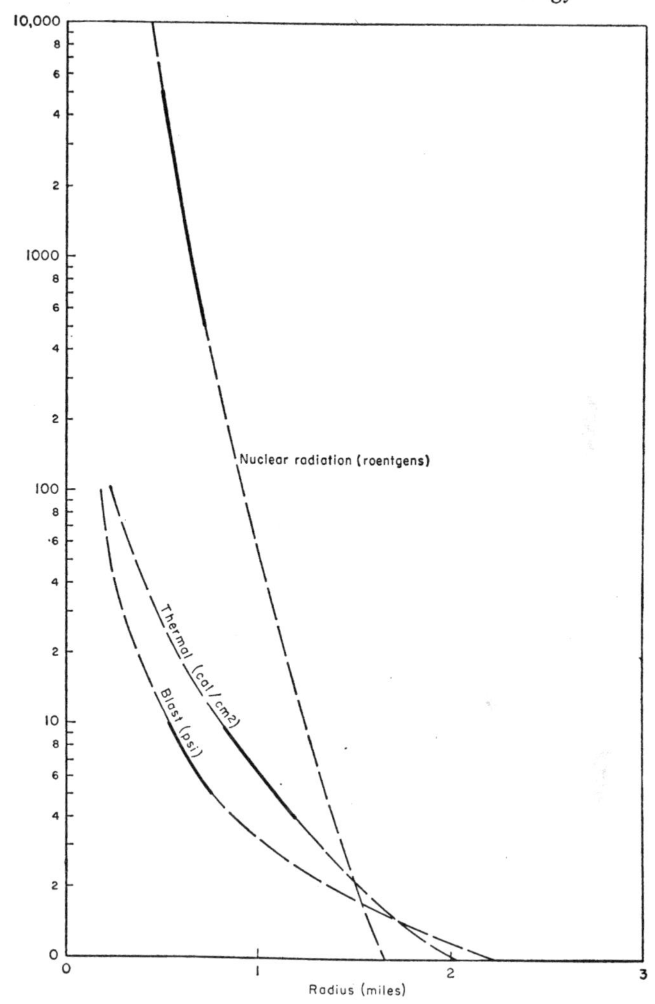

FIG. 1. Weapon effects at surface (20 KT)

Second, thermal radiation, as usually defined, also disappears. There is no longer any air for the blast wave to heat, and much higher frequency radiation is emitted from the weapon itself.

Third, in the absence of the atmosphere, nuclear radiation will suffer no physical attenuation and the only loss in intensity will arise from reduction with distance. As a result the range of significant (that is, harmful) dosages will be many times greater than is the case at sea level.

Figure 2 shows the dosage-distance relationship for a 20-kiloton explosion when the burst takes place at sea level and when the burst takes place in space. We see that in the range of 500 to 5,000 roentgens the space distances are of the order of 10 to 20 times as large as the sea-level distances. At lower dosages the difference between the two cases becomes even larger.

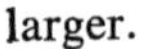

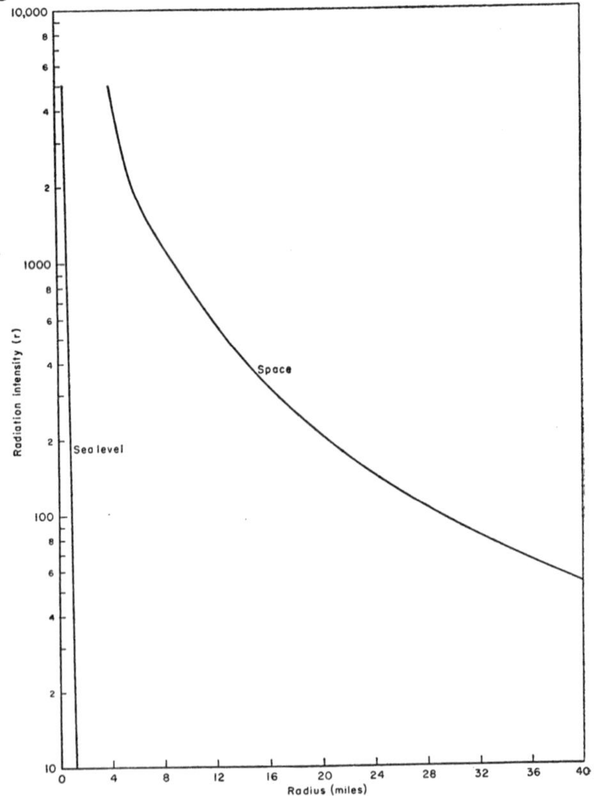

FIG. 2. Nuclear radiation intensities (20 KT)

A yield of 20 kilotons has been used here as an example to show the dominance of nuclear radiation effects in space; however, it may well be that multimegaton warheads, rather than 20-kiloton warheads, will be far more representative of space defense weapons. With such weapons the lethal distances (from nuclear radiation) in space may be perhaps hundreds of miles. The meaning of such huge lethal distances in possible future space warfare cannot now be assessed. It does seem clear, however, that, unless heavy shielding is feasible, manned space-combat vehicles will be considerably more vulnerable to nuclear defense weapons than their unmanned counterparts.

B. Possible Communication Effects

On August 1 and 12, 1958, nuclear warheads were detonated in missiles over Johnston Island in the Pacific.[2, 3] These detonations were accompanied by impressive visual displays seen over wide areas, leading observers to the opinion that the detonations took place at very high altitudes.[4-7] These displays were even seen on Samoa, some 2,000 miles from Johnston Island.

The visual displays were accompanied by disruptive effects on radio communications. Specifically, most commercial communication systems operating on the high-frequency (about 5 to 25 megacycles) bands in the Pacific noted substantial disturbances. Most links within a few hundred miles of Johnston Island experienced "outages" for as long as several hours, at various times over a period of about a day. In general, the effects on high-frequency communication links appear to have been quite similar to the effects produced by giant solar flares.

Notes

[1] The Effect of Nuclear Weapons, U. S. Department of Defense, published by the Atomic Energy Commission, June 1957.

[2] Note to Editors and Correspondents, U. S. Atomic Energy Commission, Department of Defense, Joint Office of Test Information, August 1, 1958.

[3] Note to Editors and Correspondents, U. S. Atomic Energy Commission, Department of Defense, Joint Office of Test Information, August 12, 1958.

[4] Atomic-Like Flash Seen Here—Nuclear Rocket Test Indicated, The Honolulu Advertiser, August 1, 1958.
[5] Samoa Bulletin, August 1, 1958.
[6] Samoa Bulletin, August 15, 1958.
[7] Cullington, A. L., A Man-Made or Artificial Aurora, Nature, vol. 182, No. 4646, November 15, 1958, p. 1365.

18

Cost Factors and Ground Facilities

A. Nature of Space Activity

For some time to come, astronautics will require the most ambitious kind of research and development activities. It is well known that research and development programs are characterized by risks and uncertainties. Research and development procurement decisions are fundamentally different from decisions to buy or not to buy established commodities. It is frequently necessary to fund several alternative developments in order to hedge against uncertainties. The fact must be accepted that it is not obvious that many good development decisions today will guarantee a final product. Of course, resource limitations make it necessary to select programs wisely to insure that the most promising alternatives are pursued.

In addition, the high cost of launching payloads into space, even when using basic components already perfected as part of the military missile programs, should be recognized.

B. Inheritance from Weapon System Programs

In the early years, at least, of space flight development, the carry-over from military weapon system programs will be the nation's chief set of assets in astronautics. Generally, a large amount of money can be saved by taking space flight hardware from advanced points on the production lines of ballistic missile hardware wherever possible. It should be noted that the Explorer, Thor-Able, and Juno II programs have made substantial use of existing missile hardware.

In this connection it should be noted that if space flight programs are carried on in parallel with the ballistic missile effort and use the organization and facilities of that effort, a certain degree of interaction is bound to occur, some conflicting, some mutually supporting. The total cost of the space flight activity is then not easily isolated. To determine its full cost, proration of a number of cost elements between both space and ballistic missile programs must be considered.

C. Costs of High Priority and Uneven Workload

Another important type of cost is associated with superpriority effort. Certain of these costs are readily visible, others are quite well hidden. Such items as overtime operation, travel expenditures for furthering "crash" efforts, purchase of duplicate backup equipment, etc., can easily be discerned. However, costs incurred because of sacrificing other work or pushing alternative activity to a lower priority position are not readily evaluated.

The scheduling of a constant workload in the missile and space-flight field is inherently difficult. Time is required for preparing a major weapon for test, and scheduling of subsequent tests must usually await the results of a previous testing operation. Moreover, the testing operation itself, both before and during missile countdown, is fraught with delay. And many space activities must wait upon the appropriate relative positions of celestial bodies in their orbits and upon desirable weather conditions.

D. Cost Trends of the Missile Era

Beyond these general observations on cost sensitivities it is possible to detect some specific trends which will probably be carried over from the missile field into astronautics:

(*a*) There is a trend toward concentration of highly trained personnel, purchase of expensive test equipment that is used infrequently, and construction of extensive development facilities. These things are costly.

(*b*) A second trend of missile (in contrast to aircraft) operations is the growth, both absolute and relative, of ground-support equipment. Flight hardware in missile operation does not occupy the prime cost position that it does in manned-aircraft operations.

(*c*) Third is the trend toward larger expenditure for prototype development and purchase of sizable amounts of test hardware. Cost estimation of these elements is difficult because the operation involves low volume and is of the job-lot type, and because in many cases development is performed by increasingly elaborate organizational complexes made up of many contractors. This is necessary in order to acquire diversified technical skills and because of the magnitude of the programs involved. Since financial systems vary from company to company, costs are hard to estimate.

(*d*) Finally, there is a trend toward using more components requiring individual handling in manufacture, with the actual fabrication processes absorbing a reduced percentage of the total cost. The sums required for assembly, calibration, and testing operations have increased substantially, but are much less readily estimated. Use of such rudimentary estimating methods as direct manufacturing labor hours and cost per pound alone are insufficient cost indicators for missile and space activities.

E. Sensitivity of Research and Development Costs to Flight Test Program Size

Research and development cost is highly sensitive to the size of the required flight test program. The single-shot flight test vehicles and tests to failure of components are prime characteristics of missile and space testing. Prime hardware is totally destroyed in each flight test. Many other tests, whether conducted during early development, modification, or acceptance, can also be expected to result in extensive equipment destruction. To attain a reasonably high degree of reliability, large numbers of test vehicles are necessary.

F. Financial Burden—Short Term and Long Term

Two somewhat different phases of space-flight activity are envisioned—the first covering the next 2 to 4 years, and the second running from that point into the future. During the coming 2- to 4-year period, components inherited from current military and other programs offer an efficient means, in a cost sense, for achieving such tasks as launching small satellites and space probes. At the same time, expenditures will be required for new component developments and for studying extraterrestrial environments. In addition, preliminary systems research studies and experiments on futuristic space equipment probably will be required to determine their suitability for a desired mission.

After the initial period of space research, the requirements for more hardware and more extensive facilities will probably grow substantially. One very large step will be putting man into space. The requirements of a program to achieve reasonable probability of safe launching, flight, and return will probably be extensive. With man in the space-flight vehicle, we should again expect to see flight hardware accounting for the lion's share of system cost.

In general, then, the types of major programs for which costs probably can be anticipated in the next 2 to 4 years are

1. Space probes.
2. Scientific and reconnaissance satellites.
3. New component developments, including studies of basic materials.
4. Studies of the space environment and extraterrestrial bodies.
5. Studies and experiments on potential futuristic hardware.
6. Man in space, X-15, Dyna-Soar and related activities.

Emphasis in later years will probably be on more hardware and more launch and tracking facilities to meet the requirements of manned travel to the Moon and the other planets, the assembly of space stations, the construction of a Moon base, etc.

G. Flight Vehicles

A sizable, although not major, portion of the total contractor bill is likely to be expenditures for complete research and development hardware systems. Flight hardware manufacturing costs are often estimated on a cost-per-pound basis, and although this simple statistic has weaknesses, it probably will continue in wide usage. There is no strong reason to believe that it should not be applied to estimating the costs of satellites or space stations just as it is in the aircraft and missile field.

In general, it can be stated that as total production volume increases, the cost per pound of equipment declines.

It is also true that when the function of a particular hardware item remains constant—in other words, no increase or decrease in instrumentation, no requirement for new types of materials—then the cost per pound for the system will decline as the size of the vehicle increases.

In the missile area the percentage of total hardware cost consumed by each of the missile's components varies with the function of the missile; but as an example, for a liquid-propellant ballistic missile with self-contained guidance, costs can be broken up on an approximately 20-30-20-20-10 basis among structure, controls and subsytems, propulsion, guidance, and payload container.

On the basis of fragmentary cost and design feasibility information, it appears that *high-energy propellants* will become more attractive than liquid oxygen-kerosene for an escape-velocity vehicle when the payload exceeds something like 30,000 pounds. Despite their high cost per pound of dry weight and per pound of propellant, fluorine-hydrazine or nuclear-propelled rockets will display lower total system costs in this high-payload class. For large payloads, their smaller volumes and weights will be sufficient to compensate for their higher per pound totals.

H. Ground Equipment and Facilities

Another cost area of great importance is ground equipment and facilities. This includes not only launch and guidance facilities of the kind currently in existence at Cape Canaveral [1] and the Pacific Missile Range, but also the future requirement for facilities throughout the world for tracking, observation, communication, recovery, etc. These sites may individually be relatively small and inexpensive, but they may also be numerous and located in out-of-the-way places.

Some form of ground tracking—radio, infrared, and/or optical—will be required by all space missions in order that their trajectories may be observed and monitored from the Earth. The rotation of the Earth, and the nature of space vehicle trajectories, make it necessary that tracking stations be located literally all around the globe. The ground complex set up as part of the Vanguard effort includes a string of optical tracking stations reaching south to Chile and a radio-site fence stretching around the world latitudinally. These stations are potentially permanent astronautical assets, and will have to be enlarged and supplemented for further space activities.

Ground facilities will be required for control, landing, and recovery of returning space vehicles. These are obvious and important parts of manned space flight operations. Unmanned systems will also pose recovery problems—for example, returning circumlunar vehicles containing photographic film or other experimental material. Uncertainties in the point of return lead to a need for some form of search capability to cover large areas of land and water.

Major functions that must be performed at a launching facility include final assembly and test of the flight hardware and associated equipment; operation of the test range—both the routine functions of maintenance and supply of the range and the actual operation of the launch and control network; processing of the data received from flights; recording and mathematical treatment of radio and radar data; as well as sizable amounts of photographic work.

The question of launch facilities is also influenced by the fact that certain geographical locations are demanded for achieving certain trajectories and dealing with the hazards associated with large rocket operations. For safety reasons, comparatively isolated sites are required.

The launching of rockets with payloads of 20,000 to 100,000 pounds or more is being discussed. The propellant loads of such rockets may be millions of pounds of energetic chemicals, raising some rather considerable safety problems. The minimum requirement is a great deal of open space around the launch site.

Launching facilities are inherently expensive and suitable locations are not plentiful. The possibility of constructing artificial launching islands is an interesting alternative. Location of such islands a few miles offshore might prove a relatively inexpensive method of providing launch facilities for large, potentially hazardous systems.

Launch bases also pose an additional geographical problem in that they require hundreds or thousands of miles of range over ocean or lightly inhabited areas. At the same time they require a chain of stations for flight monitoring. The farther such bases are from their source of industrial and military support, the more expensive they become to operate.

Notes

[1] Information Guide—Air Force Missile Test Center, Patrick Air Force Base, Air Research and Development Command.

19

Current Programs

The detailed characteristics and capabilities of the rockets associated with the United States military ballistic missile program are of necessity classified information. However, a general assessment of the satellite and space-flight capabilities of these missiles can be drawn from information made public in official releases and from the basic principles of rocket flight. Large variations in the performance figures quoted for a given missile may be expected because various components may be combined in many ways to exploit the potential of the launching vehicle.

A. Vanguard

The Vanguard satellite vehicle is currently the only known system that was designed explicitly as a satellite launcher. This system and its flight activities have been quite completely described in the open literature since its inception in the summer of 1955.[1] The three rocket-powered stages combine the characteristics of the basic types of rocket powerplant design —turbopump-liquid, pressurized-liquid, and solid propellant— in a single vehicle weighing about 22,000 pounds with an initial thrust of 28,000 pounds. It is designed to place into a satellite orbit some 20 pounds of payload and about 55 pounds of third-stage casing. Many of the flight and ground components and installations originally associated with the Vanguard program have now been combined with other flight vehicles to further the exploration of space.

B. Redstone

The Redstone, developed by the Army Ballistic Missile Agency and now in service use, is a surface-to-surface missile with a range of about 175 nautical miles.[2, 3] The thrust of the Redstone is about 75,000 pounds,[4] indicating a gross weight in the 40,000- to 50,000-pound class. The original propellant combination of liquid oxygen and ethyl alcohol was recently modified to include *hydyne* as the fuel (replacing the alcohol) and giving an increase of 12 percent in the missile range.[5] This improved fuel was used in the successful Explorer I satellite launching on January 31, 1958, and all subsequent launchings in the Explorer series. The complete satellite launching rocket used a modified Redstone (lengthened tank section) in combination with three upper stages of solid-propellant rockets. The Explorer placed about 31 pounds on orbit with 18 pounds of instrumented payload.[6, 7]

C. Thor and Jupiter

The Thor (Air Force) and Jupiter (Army) intermediate-range ballistic missiles (IRBM's) are currently in an advanced stage of development. Both missiles are designed for a range of 1,500 miles and have a reported thrust of about 150,000 pounds and a gross weight in the 100,000-pound class.[8] The similar capacities of Thor and Jupiter to serve as launching vehicles for space flight are being exploited in the Pioneer Moon rocket program.

The Thor booster has been combined with the second-stage rocket of the Vanguard to form a two-stage assembly designated "Thor-Able." This vehicle, carrying a test re-entry nose cone, traversed a range of about 5,500 miles in July 1958, and is the first ballistic missile known to have achieved this feat.[9]

The Thor-Able, with a small solid-propellant third stage, is the launching rocket for one type of Pioneer Moon vehicle. A total payload weight of about 85 pounds can be sent on a trajectory to the Moon,[10] and somewhat less on an interplanetary trajectory. On the basis of this lunar-shot performance, it can be estimated that the Thor-Able as a satellite launcher could probably place payloads of 300 to 500 pounds in a satellite orbit.

The Thor, combined with a second-stage vehicle developed by Lockheed Aircraft, using a Bell-Hustler liquid-propellant engine, is the first satellite launching vehicle for Project Discoverer. This combination is to place in orbit a total weight of about 1,300 pounds, including the burnout weight of the second stage. The usable payload in orbit is to be some hundreds of pounds, and some payloads are to be physically returned to Earth for recovery.[11, 12]

The Jupiter IRBM has also been combined with an upper-stage assembly (a solid-rocket cluster similar to that used on Redstone-Explorer series) for lunar flights. This launching vehicle, called Juno II, can send a Pioneer payload weighing about 13 pounds to the vicinity of the Moon.[13, 14] The Juno II is also to be used to launch Earth satellites, including a 100-foot inflatable sphere.[15]

With improved upper stages, Thor and Jupiter can launch satellites weighing as much as 2,000 pounds.[16]

The approximate cost of a Thor rocket is about $1 million.[17]

D. Atlas and Titan

The Atlas and Titan intercontinental ballistic missiles (ICBM's) (Air Force) are designed to deliver a thermonuclear warhead to a range of about 5,500 miles.[18] The Atlas is currently in flight test, with operational units to be deployed late in 1959.[19] The Titan is approaching flight-test status. The gross weight of both missiles is in the 200,000-pound class, with a takeoff thrust in excess of 300,000 pounds. The great power of the Atlas and Titan missiles can be exploited to place large weights on a satellite and space-flight trajectory.

The Atlas has the ability, without its military load, to place itself, tanks and all, in a satellite orbit.[20] An Atlas vehicle was placed in orbit on December 18, 1958, carrying some 150 pounds of payload at an average orbital altitude of 500 miles.

Unofficial announcements concerning the NASA man-in-space program indicate that the Atlas or Titan can place a payload of about 2,500 pounds into a satellite orbit at an altitude of 150 miles.

If the Atlas is combined with an upper stage weighing some 20,000 to 30,000 pounds and using high-energy propellants, a payload of 6,000 to 8,000 pounds can be placed in a satellite orbit at an altitude of about 400 miles.[21] A high-energy upper stage, of unannounced size, for use on Atlas, has been placed in development, and the resulting composite vehicle is expected to be able to launch satellite payloads of 8,000 to 10,000 pounds.[22, 23]

E. Large-Engine Developments

Developments of rocket engines in the 1.5-million-pound thrust class have been initiated.[24, 25] A single engine of this approximate thrust, combined with a suitable vehicle, could probably place a payload of roughly 30,000 to 50,000 pounds into a satellite orbit at an altitude of 300 miles, and many thousands of pounds on lunar or interplanetary trajectories.

Nuclear rocket development under Project Rover is expected to lead to very large payload capabilities.[26, 27]

For a summary of current and expected payload-weight capabilities, see table 1.

TABLE 1.—*Summary of current and expected payload-weight capabilities*

[All payload weights shown in pounds]

Item	Vanguard	Jupiter-C	Thor/Jupiter	Atlas/Titan	(?)	(?)
Weight pounds	22,000	40,000-50,000	~100,000	~200,000	—	—
Thrust do	28,000	75,000	~150,000	~300,000	1,500,000	3,000,000
300-mile satellite	20	15-30	100-2,000	2,000-8,000	37,000	75,000
22,400-mile (24-hour) stationary satellite	—	—	25-600	500-2,500	11,000	22,000
Moon impact (circum-lunar)	—	—	50-800	600-3,000	13,000	25,000
Moon satellite	—	—	15-500	300-1,500	6,000	13,000
Moon landing	—	—	10-300	200-1,000	4,000	8,000
Venus/Mars flight	—	—	25-600	500-2,500	12,000	23,000
Jupiter flight	—	—	10-400	300-1,500	2,500	5,000

NOTE.—The indicated spread in values is not intended to be associated with a specific missile but, rather, to indicate the possible capabilities with varying combinations of components.

F. X-15 Manned Research Vehicle

The X-15 manned research vehicle is being developed and tested by North American Aviation under the sponsorship of the Air Force, the Navy, and the National Aeronautics and Space Administration. This program had its beginning in research conducted by the then NACA as early as 1952; the go ahead to develop and manufacture the X-15 (three vehicles will be constructed) was given to North American in late 1955. The X-15 is designed to obtain knowledge of flight conditions at extremely high altitudes (approximately 100 miles) and at advanced flight speeds (up to 3,600 miles per hour). The vehicle itself is 50 feet long and weighs more than 31,000 pounds. Some 600 temperature- and 140 pressure-sensing devices will be carried, as well as special equipment to measure structural and aerodynamic loads and pilot reaction. An interesting feature of the X-15 is its two sets of controls: the first, aerodynamic surfaces, will provide control while the vehicle is flying within the mantle of the Earth's atmosphere; and the second, monopropellant rocket thrust units using hydrogen peroxide gas, will enable the pilot to maintain proper flight positions in the vacuum conditions outside the sensible atmosphere.

The heating characteristics of the X-15 during exit from and reentry into the atmosphere will be closely measured. Structures which will be subjected to extreme aerodynamic heating have been fabricated from the steel alloy Inconel X; other structures are made from titanium and aluminum.

The vehicle will be powered by the XLR-99 rocket engine, manufactured by the Reaction Motors Division of Thiokol Chemical Corp. This engine uses liquid oxygen and liquid ammonia as propellants and develops a thrust of 50,000 pounds.

Some outstanding features of the X-15 program are (1) extensive research and development in system components, (2) considerable attention to human-factors engineering methods, and (3) hardware production of test vehicles. The X-15 includes provisions for pilot control, so the human occupant will "drive" rather than merely "ride" the vehicle. Flight testing is scheduled to begin in February 1959 at Edwards Air Force Base.[28]

G. Dyna-Soar

The Dyna-Soar development program was initiated by the Air Force in mid-1958 after about 7 years of preliminary studies and investigations by the Air Force, NACA, and industry. The Dyna-Soar—so named because its flight is based on principles of dynamic soaring—is a rocket-boosted hypersonic glider representing an advanced stage in the development of a manned orbital vehicle for bombing and reconnaissance. The manned vehicle is to be provided with a capability for a certain amount of powered flight and lift for a controlled aircraft landing after re-entry.

As with the X-15 research vehicle, attitude (position) stabilization, atmospheric re-entry, and human factors will be major problem areas. The major contractors are Boeing Airplane Co. and an industry team formed by the Glenn L. Martin Co. and the Bell Aircraft Corp. [29-33]

Notes

[1] Fact Sheet: The Vanguard Program, Department of Defense, Office of Public Information, News Release No. 364-58, April 24, 1958.

[2] Army Marks 34th Successful Firing of Redstone Ballistic Missile, Department of Public Information, News Release No. 1135-58, November 5, 1958.

[3] Von Braun, W., Inquiry into Satellite and Missile Programs, hearings before the Preparedness Investigating Subcommittee of the Committee on Armed Services, United States Senate, 85th Cong., 1st and 2d sess., pt. I, p. 585.

[4] North American Aviation, Rocketdyne Division, news release No. NR-6.

[5] See footnote 4.

[6] Explorer I, Jet Propulsion Laboratory, California Institute of Technology, External Publication No. 461, February 28, 1958.

[7] Hibbs, A. R., Notes on Project Deal, Jet Propulsion Laboratory, California Institute of Technology, External Publication No. 471, March 14, 1958.

[8] North American Aviation, Rocketdyne Division, news release No. NR-38.

[9] Project Able Fact Sheet, Air Force Ballistic Missile Division, news release No. 58-5.

[10] Pioneer Instrumentation, Department of Defense, Office of Public Information, news release No. 984-58, October 11, 1958.

[11] Department of Defense, minutes of press conference held by Mr. Roy W. Johnson, Director, ARPA, December 3, 1958.

[12] "Project Discoverer" Satellite Program Announced by DOD, Department of Defense, Office of Public Information, news release No. 1230-58, December 3, 1958.

[13] Juno II Fact Sheet, National Aeronautics and Space Administration, fact sheet No. 3, December 1958.

[14] Instrumentation of Pioneer III, National Aeronautics and Space Administration, fact sheet No. 4, December 6, 1958.

[15] National Aeronautics and Space Administration, press conference by Dr. T. Keith Glennan and Secretary Donald A. Quarles, December 3, 1958.

[16] Johnson, R. W., address before the 13th annual meeting of the American Rocket Society, New York, November 19, 1958; Department of Defense, Office of Public Information, news release No. 1182-58, November 19, 1958.

[17] Schriever, Maj. Gen. B. A., Astronautics and Space Exploration, hearings before the Select Committee on Astronautics and Space Exploration, 85th Cong., 2d sess., on H. R. 11881, p. 670.

[18] Schriever, Maj. Gen. B. A., Air Force Ballistic Missile Programs, statement before House Armed Services Committee, 85th Cong., 2d sess., Air Force Ballistic Missile Division, news release No. 58-13, February 21, 1958.

[19] Atlas Fact Sheet, Convair Astronautics, revised October 24, 1958.

[20] Astronautics and Space Exploration, hearings before the Select Committee on Astronautics and Space Exploration, 85th Cong., 2d sess., on H. R. 11881, April 15 through May 12, 1958; K. A. Ehricke, p. 633.

[21] Astronautics and Space Exploration, hearings before the Select Committee on Astronautics and Space Exploration, 85th Cong., 2d sess., on H. R. 11881, April 15 through May 12, 1958; K. A. Ehricke, p. 620.

[22] See footnote 16.

[23] Design Development Contract Let for High-Energy Upper Stage Rocket, Department of Defense, Office of Public Information, news release No. 1134-58, November 6, 1958.

[24] See footnotes 11 and 16.

[25] North American Aviation, Rocketdyne Division, press release No. NR-31.

[26] Outer Space Propulsion by Nuclear Energy, hearings before subcommittees of the Joint Committee on Atomic Energy, Congress of the United States, 85th Cong., 2d sess., January 22, 23, and February 6, 1958.

[27] Anderson, C. P., Compilation of Materials on Space and

Astronautics, No. 2, Special Committee on Space and Astronautics, U. S. Senate, 85th Cong., 2d sess., April 14, 1958. See also Congressional Record, January 16, 1958.

[28] North American Aviation, X-15 Air Vehicle Press Information.

[29] Air Force Announces Development Plans for Dyna-Soar Boost-Glide Aircraft, Department of Defense, Office of Public Information, news release No. 573-58, June 16, 1958.

[30] The National Space Program, report of the Select Committee on Astronautics and Space Exploration, 85th Cong., 2d sess., p. 7.

[31] Glenn L. Martin Co., advertisements, Space/Aeronautics, vol. 30, No. 6, December 1958, p. 78.

[32] The National Space Program, report of the Select Committee on Astronautics and Space Exploration, 85th Cong., 2d sess., p. 24.

[33] Putt, Lt. Gen. D. L., Inquiry into Satellite and Missile Programs, hearings before the Preparedness Investigating Subcommittee of the Committee on Armed Services, U. S. Senate, 85th Cong., 1st and 2d sess., pt. 2, p. 2031.

PART 3

APPLICATIONS

20

Specific Flight Possibilities

Certain kinds of space missions are particularly interesting because they can, with reasonable certainty, be accomplished with equipments and techniques available now or in the next few years. These are briefly described in the following sections.

A. Ballistic Missiles

The theoretical projection velocity required to propel a ballistic missile to various ranges is shown in figure 1. The velocity corresponds to minimum-energy trajectories fired in a zero-drag environment around a non-rotating Earth. The projection velocity approaches the Earth-surface value of circular orbital velocity as the ballistic range approaches half the circumference of the Earth—10,000 nautical miles. This theoretical velocity remains constant at that value for ranges greater than half-way around the Earth.

Trajectories for actual missiles to be fired at ranges which are more than half-way around the Earth would have to rise above the atmosphere, so that the required velocity would be slightly greater than shown in figure 1.

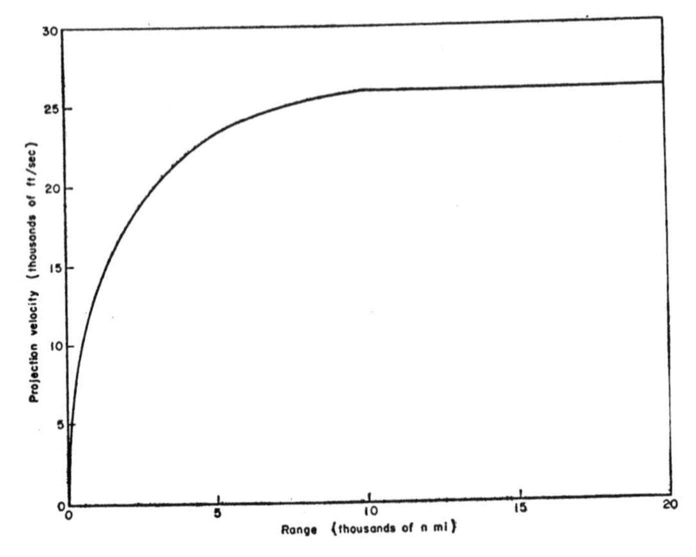

FIG. 1. Ballistic missile range

B. Sounding Rockets

The first scientific use of rockets was for vertical ascents or "soundings" of the atmosphere. The initial velocity required at the surface of the Earth for drag-free flight to various altitudes is shown in figure 2. The total time for flight in a vacuum is also shown. The total velocity required to establish vehicles in circular orbits is also included for comparison. The elliptical ascent trajectory leading to these satellite orbits is assumed to be tangential to both the surface of the Earth and to the final circular orbit. The total velocity is defined as the sum of the initial velocity required to place the vehicle on the ascent ellipse and the final velocity increment required to establish the circular orbit.

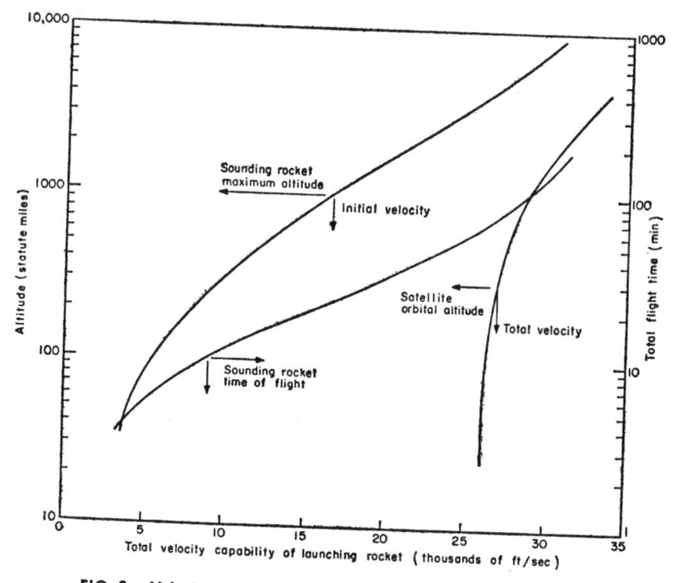

FIG. 2. Velocity requirements for sounding rockets and satellites

C. Earth Satellites

To establish a vehicle in a satellite orbit around the Earth, it is generally necessary to have two phases of powered flight, separated by an interval of coasting flight. The first powered phase is like that of a ballistic missile. After the vehicle coasts to its maximum altitude (apogee), another powered phase must give the vehicle enough speed, properly directed, to keep it in orbit (figure 3). The number of powered stages required in each of these two phases is determined by the design of the vehicle. For example, the Army's Explorer satellites were established by a single booster—the Jupiter C—firing during the first phase followed by three stages fired in succession at orbital altitude. The Vanguard satellite, on the other hand, uses two stages during the initial boost phase, and only one stage to provide the velocity needed at orbital altitude.

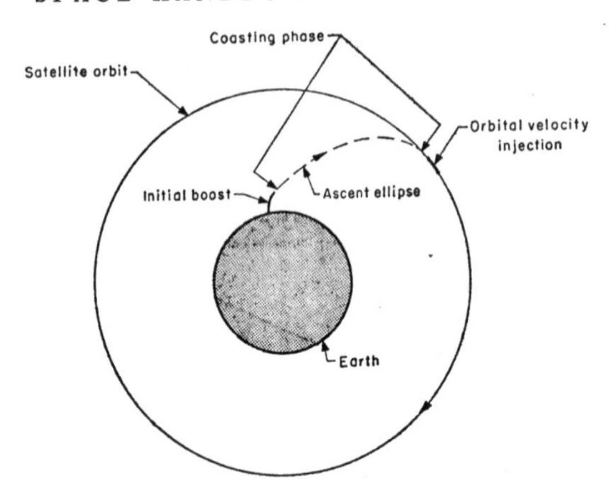

FIG. 3. Typical satellite ascent trajectory

If the designer so chooses, a satellite could also be established using a continuously powered ascent trajectory—that is, one in which the coasting phase vanishes. The Atlas satellite, launched in December 1958, followed an ascent trajectory of this type.

The main orbit features of Earth satellites are the period of revolution and the orbital velocity. These vary according to the satellite's mean distance from the center of the Earth (figure 4).

At an altitude of about 22,000 miles the period of the satellite is exactly one day and the orbital velocity is about 10,000 feet per second. If a vehicle were established in an equatorial orbit at this altitude, moving eastward with the Earth's rotation, it would remain fixed over a point on the surface of the Earth.

The equatorial bulge of the Earth and other irregularities cause disturbances of satellite orbits over long periods of time—in particular, changes in the orientation of the orbit in space.[1] These disturbances can be exploited in some applications. For example, if a satellite is established in an orbit inclined about 83° from the Equator and moving in a westerly direction, the Earth's bulge will cause the orbit plane to move in such a way that it remains fixed relative to the Sun as the Earth moves in its orbit.

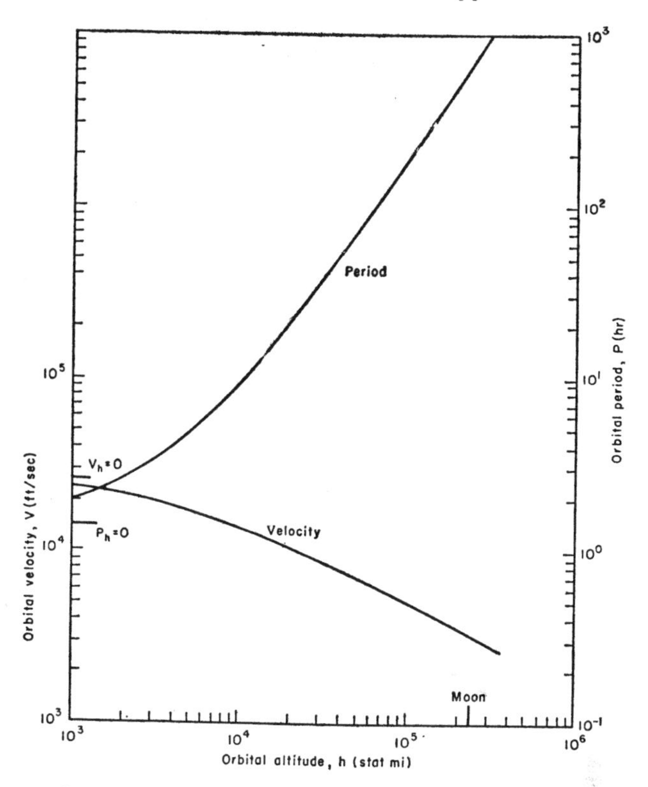

FIG. 4. Earth satellite velocity and period versus orbital altitude

An interesting extension of the basic method of launching a satellite is to provide the vehicle with an additional powered stage for use in the so-called "kick-in-the-apogee" type of maneuver (figure 5). Assume for example, that a given booster vehicle can carry a satellite to an altitude of 200 miles and that the satellite's final burnout velocity can be made greater than the circular orbital velocity required at that altitude. The resulting satellite orbit will, therefore, have an apogee altitude greater than the initial 200-mile altitude—say, 500 miles as an example. If the additional powered stage, correctly oriented, is fired at the 500-mile apogee point, the perigee (minimum) altitude of the new satellite orbit will be raised so that a final orbit can be achieved that is not limited by the projection altitude limit of the basic booster rocket.

An extra rocket stage may also be used to correct or modify the altitude or eccentricity of a satellite orbit to tailor it for a specific application. In addition, rockets can be used to change the inclination of the orbital plane relative to the Earth's Equator.

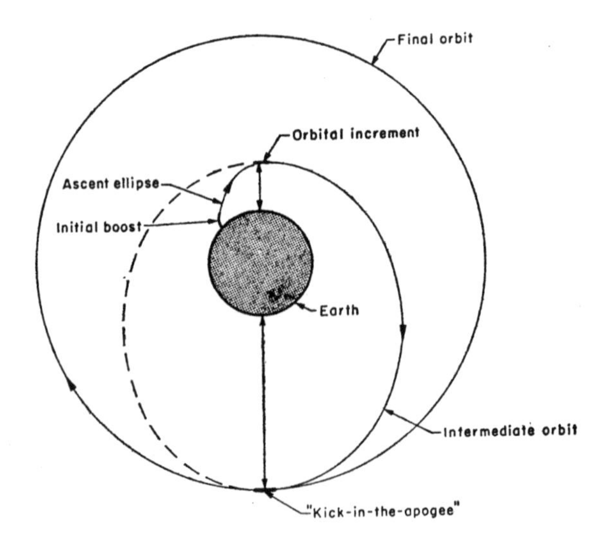

FIG. 5. "Kick-in-the-apogee" technique of satellite launching

D. Recoverable Satellites

Satellites can fairly easily be made to survive the heating encountered in returning to the Earth after natural orbit decay due to air drag in the outer reaches of the Earth's atmosphere. All that a satellite like Vanguard would need would be a shell of high-temperature alloy such as is used in jet-engine turbines.[2]

Using re-entry techniques derived from ballistic missile developments, it should be possible to bring satellite payloads back to the Earth for study. In fact, controlled recovery of biomedical and other experiments from satellite orbits is planned to start early in 1959 as part of Project Discoverer.[3]

To displace an object from a satellite orbit and bring it back to Earth, a rocket is needed on board to provide thrust at the proper time and in the proper direction. The minimum velocity that will accomplish recovery is shown in figure 6 for a range of satellite altitudes. The required ejection velocity is in no case as much as 5,000 feet per second. Vehicle re-entry velocities are also shown in figure 6.

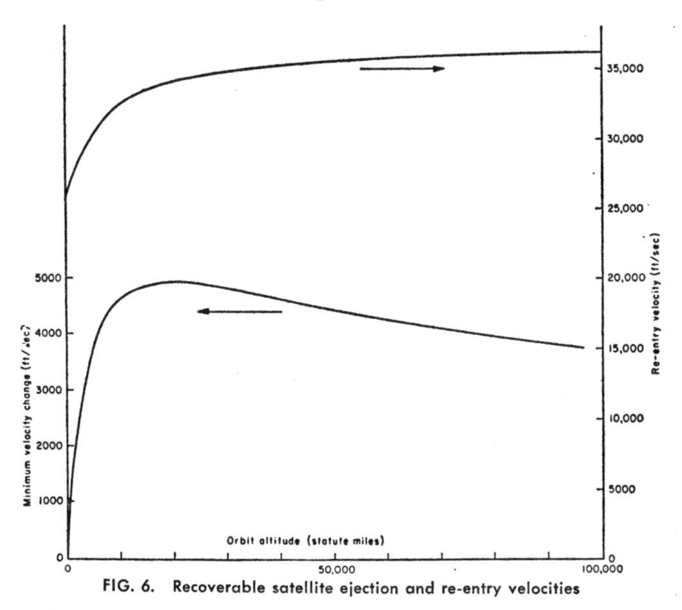

FIG. 6. Recoverable satellite ejection and re-entry velocities

If the satellite payload is to be returned to a specific recovery area, the satellite must descend fairly steeply. Such descents require greater ejection velocity than the minimum amounts shown in figure 6.

The ejection velocity required to bring down a payload from a 300-mile orbit is plotted in figure 7 as a function of the *descent range*—the distance over the ground covered by the payload from ejection to impact. The ejection velocity grows rapidly as descent range decreases.

The way in which impact location changes due to errors in control of the ejection velocity is also plotted in figure 7. Accuracy of recovery improves as the descent range becomes less. For example, at a descent range of 5,500 nautical miles an impact miss of about 4 nautical miles will be produced by an error of 1 foot per second in ejection velocity. Other sources of error, of course, also affect recovery accuracy, particularly control of the direction of ejection velocity.[4]

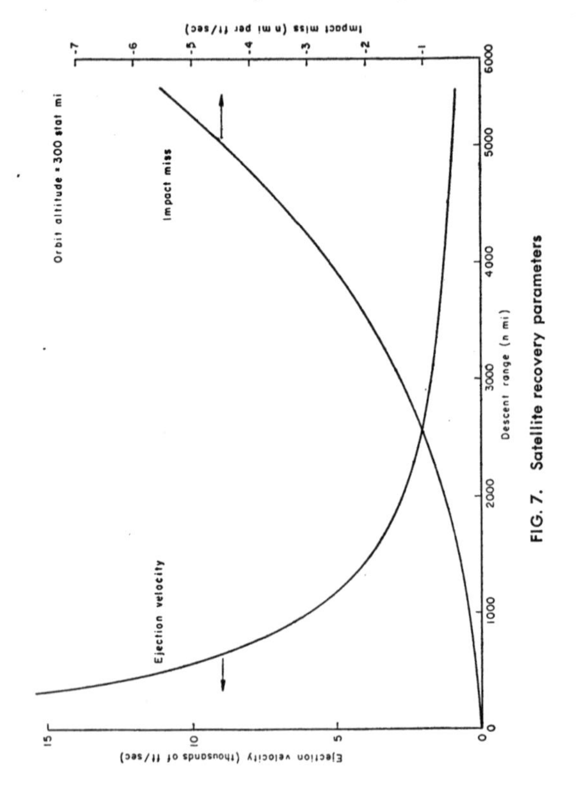

FIG. 7. Satellite recovery parameters

E. Lunar Flights

LUNAR IMPACT

The minimum velocity that will cause a vehicle to reach the Moon is only slightly less than the local escape velocity from the Earth, which at an altitude of 350 statute miles is 35,160 feet per second. An altitude of about 350 statute miles is a reasonably representative altitude at which to start free, or unpowered, flight, and it will be assumed to apply throughout as a standard reference case. The powered-flight trajectory up to this altitude is not of particular interest in this discussion.

The free-flight portion of a sample trajectory which terminates on the Moon is shown in figure 8. The trajectory is plotted in such a way that it appears about as it would to an observer at the North Pole. The elliptical nature of the trajectory over about the first 2 days of flight is evident. The effect of the Moon's gravitational attraction becomes large only during the last half day of the flight, as it causes curvature of the terminal portion of the path, leading to an impact on the surface of the Moon.

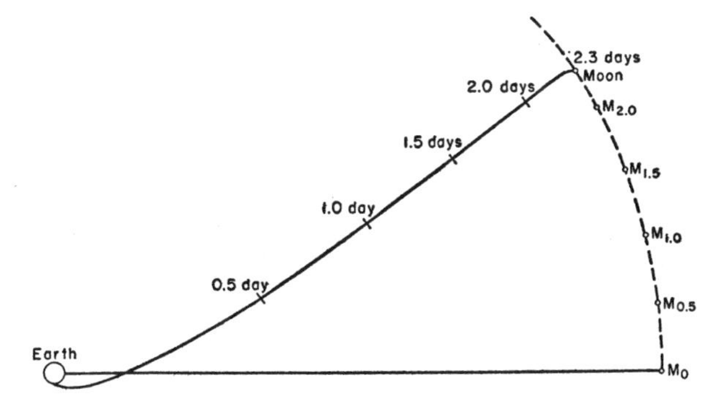

FIG. 8. Moon-rocket transit trajectory—impact

The same trajectory is shown in figure 9 as it would appear to an observer on the Moon.[5]

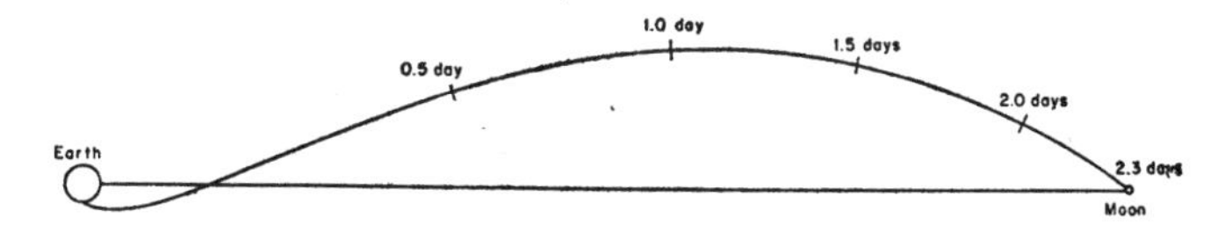

FIG. 9. Moon-rocket trajectory as seen from the Moon

The variation of the impact point on the Moon caused by errors in the initial velocity is shown in figure 10. The central curve, defined by an initial velocity of 35,000 feet per second, corresponds to the trajectory shown previously in figure 8. For a velocity that is 35 feet per second too low the impact point moves to the western limb of the Moon; for 50 feet per second too much velocity it moves to a point beyond the eastern limb. Thus, some of the possible impact points would not be visible from the Earth. It can be seen that errors of only -35 to $+50$ feet per second are allowable around this design point for simple lunar impact trajectories. This tolerance is critically dependent upon the nominal initial velocity selected.

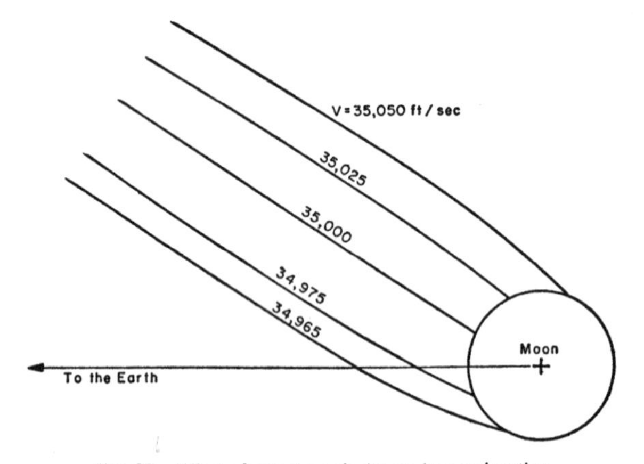

FIG. 10. Effect of varying velocity on impact location

Figure 11 presents a typical "hit band" showing allowable variations of initial velocity and path angle for trajectories which will result in impact somewhere on the Moon.

Initial velocity-path angle combinations lying along the left edge of the shaded band define trajectories which are tangent to the Moon near the eastern limb, while combinations along the right-hand edge of the band define trajectories which are tangent near the western limb of the Moon.

A curve defining the trajectories which are perpendicular to the lunar surface lies approximately midway between the solid lines.

The terminal trajectory variation shown previously in figure 10 corresponds to a vertical slice through the lower portion of the hit band.

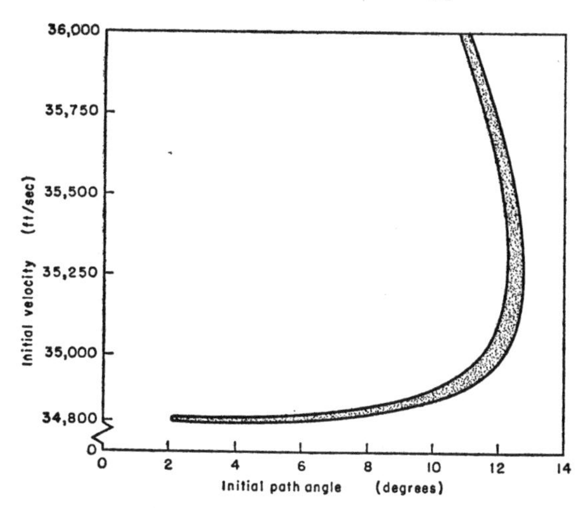

FIG. 11. Initial conditions for lunar impact

Since the slope of the hit band varies from nearly horizontal (variation in path angle) to nearly vertical (variation in velocity), the relative values of the tolerances in velocity and path angle to hit the Moon will vary as a function of the magnitude of the initial velocity.

We can define the total tolerances in either velocity or path angles as the allowable increment in the parameter which causes the impact point to shift from one limb of the Moon to the other. Plotting the tolerances as a function of the initial velocity which corresponds to an impact normal to the surface of the Moon gives the curves shown in figure 12. At near-minimum initial velocities the allowable path angle tolerance is about 4½°, but decreases rapidly to a value of approximately one-half degree for higher velocities.

The velocity tolerance rises gradually to a value of about 300 feet per second as the initial velocity is increased to a value of about 35,100 feet per second. The notation (W-E) indicates that the impact points shift from the western limb of the Moon to the eastern limb (see figure 10). Above an initial velocity of about 35,500 feet per second the velocity tolerance decreases gradually. In this portion of the curve the impacts move from the east limb to the west limb. The portion of the curve labeled (W-E-W) corresponds to the nearly vertical portion of the hit band of figure 11. In this region of the curve, the lunar impact point shifts from the west limb to a point near the east limb and then back to the west limb. The tolerance of 600 feet per second occurs for trajectories which are defined by a path angle corresponding to the left-hand edge of the "hit band." Therefore, nominal trajectories which allow these extremely large initial velocity tolerances are extremely sensitive to the exact value of the initial path angle.

The variation of initial velocity and path angle tolerances shown in figure 12 is representative of transit trajectories which are defined by relatively low initial path angles. As the initial path angle of the free-flight trajectory approaches 90° (fired from an initial position which "leads" the Moon), the reflex portion of the "hit band" in figure 11 moves to extremely high initial velocities so that the velocity of tolerance curve rises more gradually than shown in figure 12.[6]

In general, therefore, it is possible to "tailor" the nominal

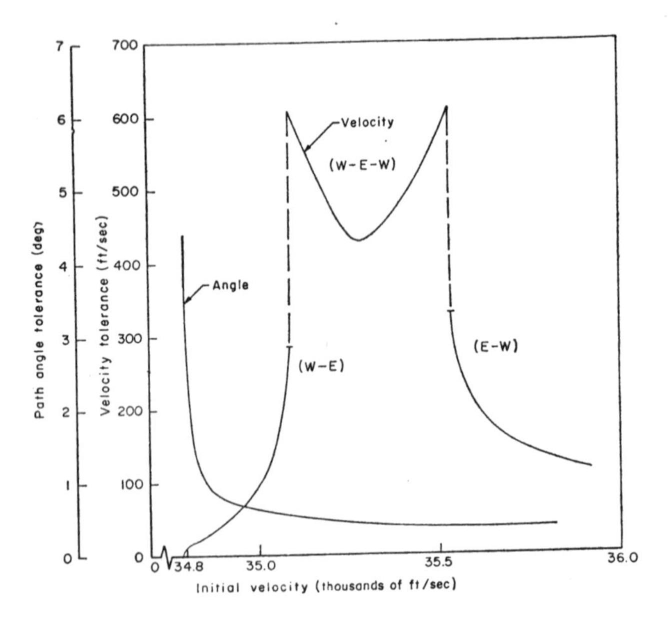

FIG. 12. Total velocity and path angle tolerances to hit the Moon

transit trajectory design point (and the corresponding initial tolerances) to accommodate particular accuracy capabilities of the ascent guidance system.

The transit time for free-flight trajectories from the Earth to the Moon is also strongly dependent upon the magnitude of the initial velocity, as shown in figure 13. The longest flight time, 5.5 days, corresponds to the minimum-velocity trajectory. If the initial velocity is increased by only 1 percent, the transit time is decreased to a value of 2 days. The flight time planned for the Air Force Pioneer Moon shots was about 2.6 days, while that planned for the Army Juno II Moon shots was only 1.4 days. From figure 13 it can be seen that this sharp difference in flight time involves a velocity difference of only about 2.5 percent.

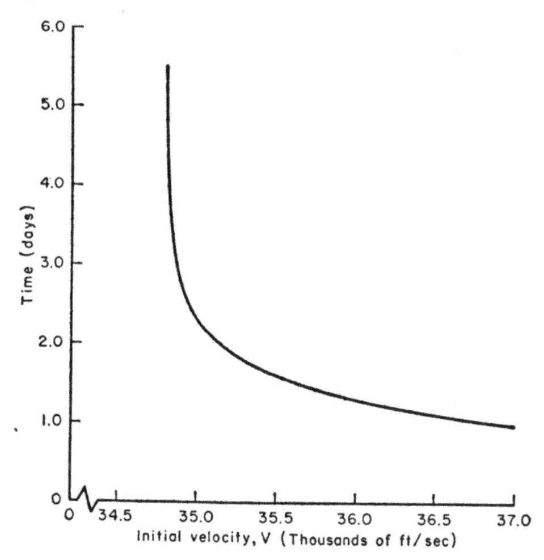

FIG. 13. Transit time from Earth to Moon

Impact on the Moon might be signaled to Earth observers by detonating about 10 pounds of standard illuminant powder.[7]

CIRCUMLUNAR FLIGHT

A vehicle can be fired on a circumlunar trajectory which returns to the vicinity of the Earth with no further propulsion stage if the initial velocity at the Earth is less than the local escape velocity.[8] The vehicle must be launched so that it will intersect the Moon's orbit at a point ahead of the Moon, in order to be swung around by the Moon's gravitational field for return.

The distance of closest approach to the Moon will vary widely for trajectories based on different initial conditions, and this distance may vary from a grazing passage to as much as 80,000 miles.

The time of closest approach to the Moon is primarily dependent on the initial velocity at the Earth, as shown in figure 13, but is also affected slightly by the distance of closest approach to the Moon. The position, on the lunar surface, below the point of closest passage is also variable.[9]

Typical examples of the two major classes of free-flight circumlunar trajectories are shown in figures 14 and 15. A typical low-velocity trajectory is given in figure 14—an example of the "figure eight" orbit. The velocity of the vehicle in the vicinity of the Moon is relatively low (but greater than lunar escape velocity), so that the Moon's motion and gravitational attraction cause the trajectory to be strongly perturbed. In this sketch, the vehicle passes above the Earth on its return. Figure 15 shows a typical trajectory defined by a somewhat higher initial velocity. In this example, the vehicle's velocity in the vicinity of the Moon is somewhat higher, so that the Moon's perturbing influence is less pronounced. The vehicle continues to some distance beyond the Moon's orbit before returning to the vicinity of the Earth. It passes below the Earth on return. Free-flight trajectories defined by velocities between those of figures 14 and 15 will return to impact somewhere on the Earth. The distance of closest approach to the Moon in both these cases is about 3,000 miles.

The total flight time from start to return to the Earth can vary from a minimum of about 6 days to as much as a month, depending upon the initial velocity. A small change in initial velocity can produce a very great change in total flight time. This is a matter of considerable practical importance, since points on the Earth are moving rapidly due to the rotation of the Earth. Thus, for example, a vehicle returning from a 10-day circumlunar trip intended for recovery at, say Edwards Air Force Base, will find the base 800 miles away if it is just 1 hour off schedule.

This extreme trajectory sensitivity for a completely unpowered vehicle can, of course, be reduced by providing it with

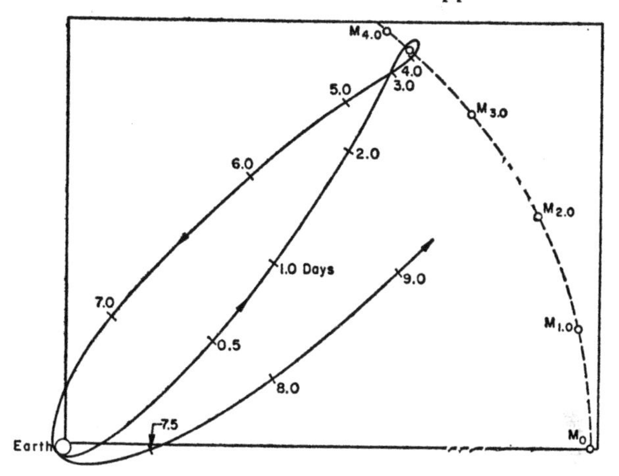

FIG. 14. Return near the Earth after passing near the Moon—returning vehicle moving in same direction as Earth's rotation

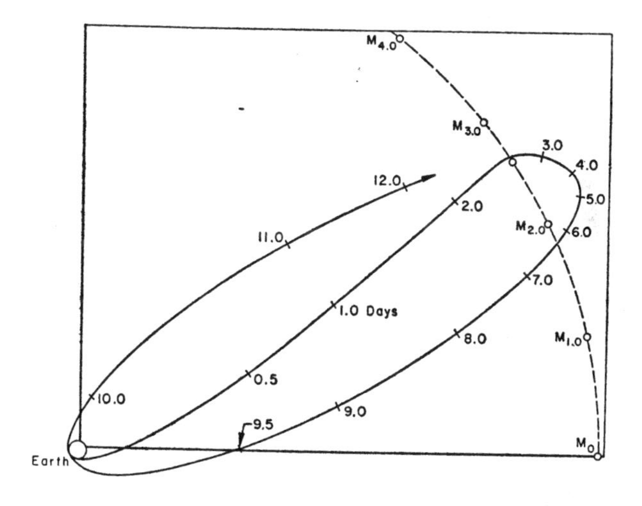

FIG. 15. Return near the Earth after passing near the Moon—returning vehicle moving in direction opposite from Earth's rotation

the ability to make one or more corrective velocity changes during transit.

A circumlunar flight with physical recovery on Earth would be suitable for such applications as acquiring high-quality photographs of the Moon's hidden side.[10] Re-entry design techniques for such a return are within the capabilities of current technology.[11, 12]

MOON-TO-ESCAPE FLIGHTS

The Moon's orbital motion and gravitational field can be used to accelerate a vehicle out of the Earth-Moon system and into an independent orbit around the Sun.

A typical trajectory which is changed from an elliptical to a hyperbolic orbit as a result of the Moon's action is shown in figure 16. The lunar-miss distance for this particular trajectory is about 600 miles, and the resulting effective initial-velocity increase is approximately 400 feet per second. The maximum possible velocity increase due to the Moon's perturbation is about 500 feet per second for a trajectory which passes just above the lunar surface.

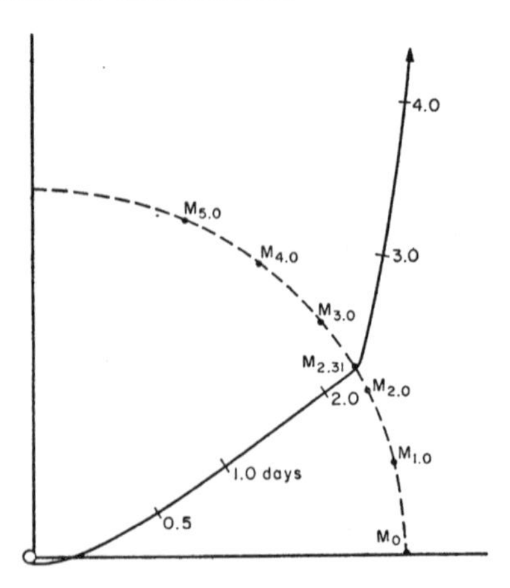

FIG. 16. Transit trajectory-escape from the Earth

LUNAR LANDING

Landing a vehicle on the Moon requires an additional rocket stage to decelerate it to a safe landing speed. The approach velocity at the Moon depends on the initial velocity at the Earth, but is never less than about 8,200 feet per second because of the gravitational attraction of the Moon. Since the lunar-escape velocity is about 7,800 feet per second at the surface, the rocket will always be on a hyperbolic trajectory relative to the Moon. Control of the direction of braking rocket thrust for landing introduces a need for control of vehicle orientation. The signal for firing the retrorocket must either be sent from the Earth or generated in the vehicle itself (for example, by measurement of the lunar altitude).

The transit trajectory for a lunar landing will be generally similar to that for a lunar impact discussed above and shown in figure 8.

LUNAR SATELLITES

Satellites of the Moon can be established if provision is made to reduce their velocity in the vicinity of the Moon.[13]

A transit trajectory for a lunar satellite is shown in figure 17. At a predetermined distance from the Moon—near the point of closest approach—the vehicle's velocity must be reduced to a value which will allow the Moon to capture the vehicle as a satellite. A representative value for this velocity change would be of the order of 4,000 feet per second—about half the value required for a lunar landing. If the velocity is reduced at the point of closest approach of the transit trajectory shown in figure 17, the initial portion of the lunar satellite will be as shown in figure 18.

A lunar satellite was the flight objective of the Pioneer shots.

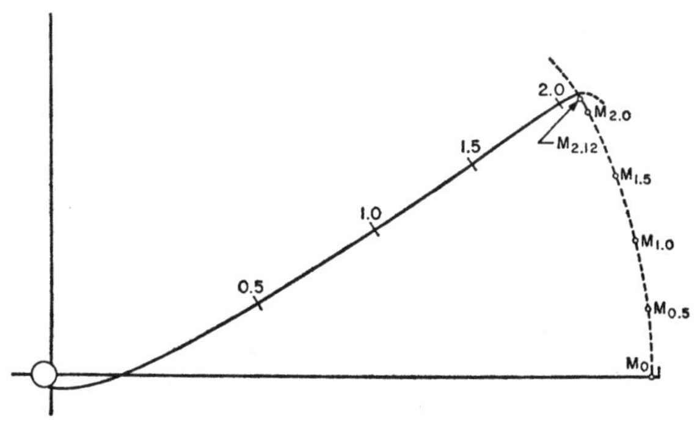

FIG. 17. Transit trajectory—lunar satellite

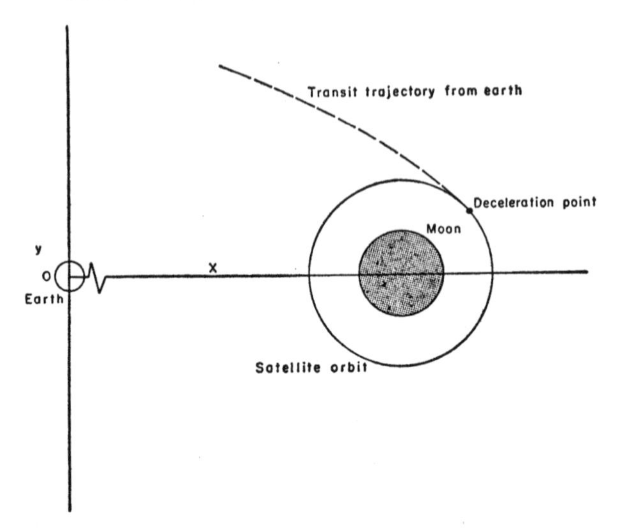

FIG. 18. Moon satellite orbit

MOON-TO-EARTH RETURN

Launching a vehicle from the Moon to return to the Earth may be of considerable interest in the future.[14] The initial velocity at the Moon would typically be around 10,000-15,000 feet per second. As in the case of Earth-to-Moon trajectories, the transit time is a function of the initial velocity. The variation is more gradual, however. An initial velocity difference of about 3,000 feet per second at the Moon corresponds to a change in transit time from a value of about 1.5 to about 2.5 days.

Merely hitting the Earth from the Moon is considerably easier than hitting the Moon from the Earth, since the diameter of the Earth is larger and its gravitational field is much stronger. Typical values of the initial tolerances at the Moon might be plus or minus 1,500 feet per second in velocity and plus or minus 5° in path angle. The requirements for hitting a given area on the Earth, however, are considerably more severe—mainly because the Earth's rotation on its axis causes the "target" to move. Thus, very accurate control of initial velocity at the Moon is required mainly to fix the flight time accurately.

SPACE BUOYS

One of the special solutions to the equations of motion in the classical "Problem of Three Bodies" predicts the existence of 5 singular points in the vicinity of two massive bodies at which a vehicle can be held stationary without application of thrust.[15] The locations of these points, known as "centers of libration," in the Earth-Moon system are indicated in figure 19. Three lie on the line joining the Earth and the Moon, while the other two form equilateral triangles with the Earth and Moon.

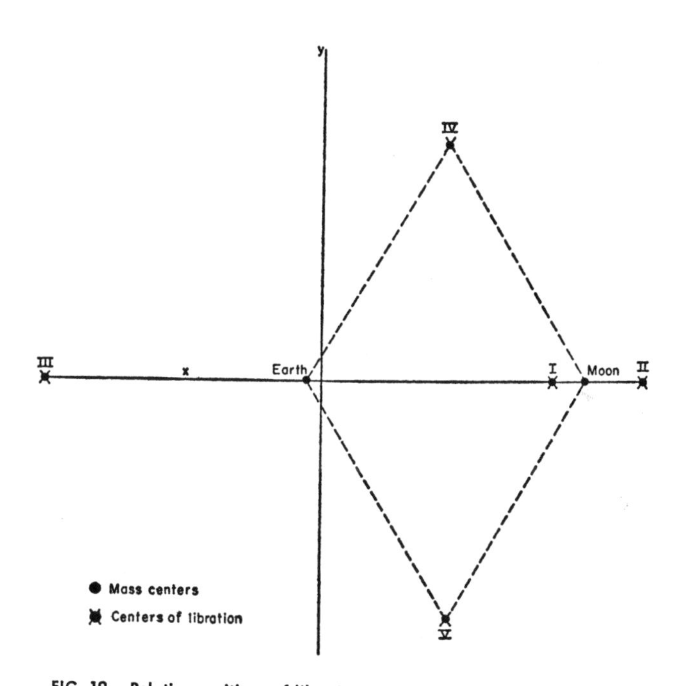

FIG. 19. Relative positions of libration centers in the Earth-Moon system

Motion in the vicinity of the straight-line points is unstable; that is, any small displacement from the exact point will result in the particle's moving indefinitely far away.

Motion of a particle in the vicinity of the equilateral triangle points, however, is stable—at least to disturbances arising from within the system—so a vehicle near one of these points could be made to "float" in space as a "space buoy." [16] The "Trojan" asteroids, which oscillate in the vicinity of such points in the Sun-Jupiter system, provide a check of this theory.[17]

If vehicles containing the appropriate equipment could be established at, or near, each of these equilateral triangle points in the Earth-Moon system, they could provide a baseline which is about 400,000 miles long for such uses as tracking vehicles on interplanetary flights. The displacement of these "space buoys" by the Sun's net gravitational attraction would have to be investigated. Also, if the bodies were low-density structures, the acceleration due to solar radiation pressure would cause displacements.

F. Interplanetary Flights

ARTIFICIAL ASTEROID

A small manmade body in orbit around the Sun would be properly termed an artificial asteroid. Such vehicles, carrying instruments, can be established by simply launching them at any speed greater than escape velocity, regardless of direction. This fact has been recognized in laying out the plan for the Army lunar probes, which will be launched with enough speed to take up an artificial asteroid orbit if they miss the Moon.[18]

Instrumented asteroids could probe the space environment around the Sun. Also, if made optically observable—by carrying large reflecting balloons—or fitted with transponders for radar range measurements, such vehicles could be used just as natural asteroids now are in obtaining a fundamental measurement of the length of the astronomical unit (distance from Earth to Sun).

PLANETARY SATELLITES

A vehicle brought near one of the planets could be placed in orbit around the planet in much the same manner as satellites of the Moon are established. Some form of terminal guidance would be required, with propulsion to make it effective.

Manmade satellites for Mars and Venus should be achievable in a relatively few years. Mercury and Jupiter could be given satellites at a later date.

LANDINGS ON VENUS AND MARS

It should be possible, with provision of midcourse and/or terminal guidance, to land some hundreds of pounds of instruments on the surfaces of Venus and Mars with basic rockets now in development.

Any flights in interplanetary space will require use of flight equipments with very long reliable life—a year or more in most cases. The actual achievement of interplanetary flight is critically dependent upon developing long-lived, reliable equipment.

G. Manned Flights

No firm assessment can be made at this time of the extent to which manned vehicles can undertake these flights (except, of course, to cull out obviously unsuitable ones like unchecked impact on the Moon). Performance calculations alone would indicate feasibility of rather extensive flights, like circumlunar flights and flights to even greater distances from the Earth. However, firm judgment on these kinds of possibilities must await results of such manned-flight experiments as are planned in the X-15 program, the NASA manned capsule satellite program,[19] and the Dyna-Soar program.

H. Effects of Time of Launching

NATURE OF THE PROBLEM

The payload that can be carried to any target region with a given launching rocket depends to a large extent upon the nature of the trajectory followed. Some relative positions of the Earth and a target body allow transit between them on favorable trajectories, while other positions impose trajectories that seriously tax rocket performance. Since the positions of bodies in the solar system are continuously changing with time, trajectory requirements change continuously, and so, therefore, does the payload that can be carried. Hence, there are literally "good days" and "bad days" for every kind of space launching—in fact, with limited rocket performance, there are some days that are simply "no-go days." Since the position of the launch point on Earth is moving with the Earth's daily rotation, there are also "good" times of day and "bad" times of day. A launching arranged for a given time on a given day must take place on schedule or not at all, unless a large sacrifice in payload is made.

LUNAR FLIGHTS

Generally speaking, a rocket system displays its best payload capability when fired over a rather closely circumscribed powered-flight trajectory; and any substantial departure from this trajectory will greatly reduce its payload. For large liquid rockets on lunar flights, the powered-flight trajectory terminates with the rocket moving nearly horizontally with respect to the Earth. The trajectory has a certain total length determined by the duration of the propulsion phase. Given these fairly inflexible powered-flight conditions, there is a corresponding limited range of positions of the Moon that are reachable from a given launch site.

The shape of the powered trajectory determines the time of day of favorable launching—10 minutes is a fairly representative figure for the duration of this favorable period, which occurs at approximately daily intervals.

The plane of the free-flight trajectory from Earth to Moon is defined by three points: the center of the Earth, the center of the Moon, and the location of the launch point. The inclination of this plane to the Earth's Equator changes as the Moon moves around the Earth in its monthly journey. A rocket fired in this plane at the "best" time of the month will be aimed more nearly east than one fired at some other time. The more nearly east a rocket is fired, the more velocity it will pick up from the Earth's rotation, about 1,000 miles per hour at the Equator, and the larger will be its payload.

Even at the most favorable time of month, a launching

from a latitude greater than 28.5° will not be directed exactly east, since the Moon's orbit never tips any more than 28.5° from the Equator. However, this tip of the Moon's orbit varies over an 18.6-year period, and at the points of maximum inclination, a Moon launching can be made more nearly east than at other times. Thus, there are best times within this 18.6-year cycle.

The distance from Earth to Moon varies somewhat during the month, but this distance variation itself does not have a large influence on the choice of launching times.

In summary, then, there is a best time to shoot during a given day, a best day during a given month, and a best month once every 18.6 years. The influence of time of day on payload is very strong (strong enough to make the whole operation possible only at the best time); the influence of the time of month is less strong (not likely to be the difference between feasibility and infeasibility except with marginal systems); and the effect of the 18.6-year cycle is rather minor.

It is of interest to note that at present the Moon is near its least favorable inclination. It will reach its best inclination in 1969.

If visual reconnaissance of some portion of the Moon's surface is desired, then the phase of the Moon will dictate the time of month of launch, even though the unfavorable position of the Moon may cause a considerable reduction in payload weight. For example, if it is decided to photograph the far side of the Moon, then the lunar rocket should pass behind the moon during the time of new Moon, when the far side is illuminated.

Another consideration that may be important in determining launch time is any restrictions on the direction of launch available due to a need to pass over a guidance range or a need to avoid passing over population centers.

INTERPLANETARY FLIGHTS

The free-flight trajectory of a vehicle fired from the Earth to another planet will generally be a section of an elliptical orbit around the Sun (figure 20). Maximum payload will be carried if the vehicle traverses exactly half of this ellipse between perihelion (point closest to Sun) and aphelion (point farthest from Sun), so the start and end of the voyage lie directly across the Sun from one another. The payload will be less if the vehicle must traverse either less than or more than half an ellipse.

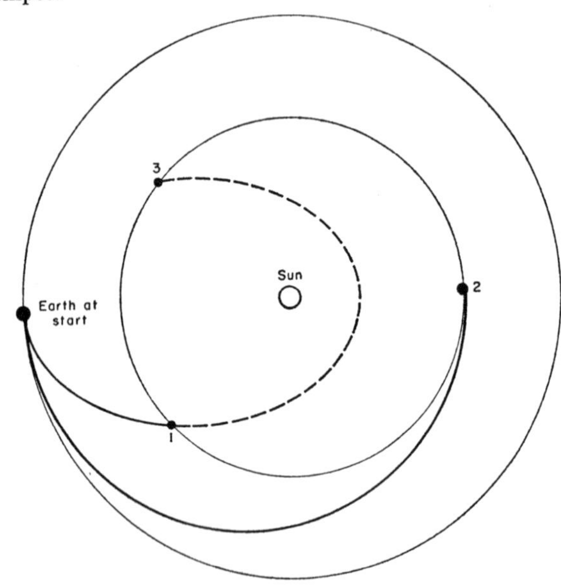

FIG. 20. Flights to Venus in various positions

The situation leading to maximum payload (or any payload at all in many cases) is only possible, of course, when the relative positions of Earth and the target planet are proper—a fairly rare occurrence except for the inner planets Venus and Mercury.

Figure 20 illustrates the situation for flights to Venus. The maximum-payload case, in which the vehicle travels over exactly half an ellipse (Venus in position 2), involves a flight time of about 4 months. The flight to Venus in position 1 takes less time, but requires greater initial velocity and therefore permits less payload for a given rocket. Flight to Venus in position 3 takes more time and more initial velocity than a flight to position 2. Thus, flights to a type-3 position have the disadvantage of both long flight times and low payload, relative to the position 2 case. Flights to positions like 1 also reduce net payload; however, the advantage of lower flight time may be more than enough to offset this disability. For example, in manned flight the weight of stores necessary to sustain the crew will be less if flight time is less, so a short flight at lesser total payload at takeoff may actually arrive at the target planet with more usable payload.

TABLE 1.—*Most favorable launching dates*

Earth to Mars	*Earth to Venus*
Oct. 1, 1960	June 8, 1959
Nov. 16, 1962	Jan. 13, 1961
Dec. 23, 1964	Aug. 16, 1962
Jan. 26, 1967	Mar. 28, 1964
Feb. 28, 1969	Oct. 27, 1965
	June 5, 1967
	Jan. 11, 1969

NOTE.—The increases in launch velocity required at times beyond these most favorable launch dates are indicated in figure 21.

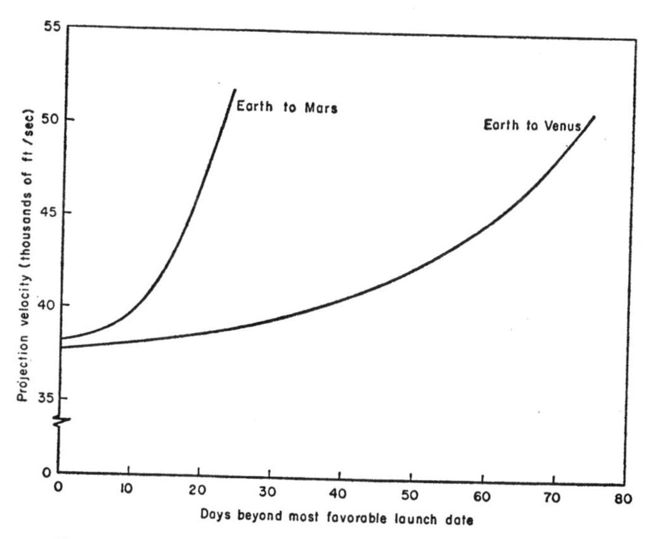

FIG. 21. Velocity penalty for launching after most favorable dates

Notes

[1] Kalensher, B. E., Equations of Motion of a Missile and a Satellite for an Oblate-Spheroidal Rotating Earth, Jet Propulsion Laboratory, California Institute of Technology, memorandum No. 20-142, April 12, 1957.

[2] Gazley, C., Jr., and D. J. Masson, A Recoverable Scientific Satellite, The RAND Corp., Paper P-958, October 5, 1956.

[3] Department of Defense, minutes of press conference held by Mr. Roy W. Johnson, Director, ARPA, December 3, 1958.

[4] Frick, R. H., Preliminary Analysis of a Satellite Recovery System, The RAND Corp., Research Memorandum RM-2264, September 19, 1958.

[5] Lieske, H. A., Lunar Instrument Carrier—Trajectory Studies, The RAND Corp., Research Memorandum RM-1728, June 4, 1956.

[6] Lieske, H. A., Lunar Trajectory Studies, The RAND Corp., Paper P-1293, February 26, 1958.

[7] Dole, S. H., Visual Detection of Light Sources On or Near the Moon, The RAND Corp., Research Memorandum RM-1900, May 24, 1957.

[8] Lieske, H. A., Circumlunar Trajectory Studies, The RAND Corp., Paper P-1441, June 25, 1958.

[9] See footnote 8.

[10] Davies, M. E., A Photographic System for Close-Up Lunar Exploration, The RAND Corp., Research Memorandum RM-2183, May 23, 1958.

[11] Gazley, C., Jr., and D. J. Masson, Recovery of Circum-Lunar Instrument Carrier, American Rocket Society, preprint No. 488-57, October 1957.

[12] Gazley, C., Jr., Deceleration and Heating of a Body Entering a Planetary Atmosphere from Space, Vistas in Astronautics, Pergamon Press, 1958.

[13] Buchheim, R. W., Artificial Satellites of the Moon, The RAND Corp., Research Memorandum RM-1941, June 14, 1956.

[14] Buchheim, R. W., and H. A. Lieske, Lunar Flight Dynamics, The RAND Corp., Paper P-1453, August 6, 1958.

[15] See footnote 14.

[16] Buchheim, R. W., Motion of a Small Body in Earth-Moon Space, The RAND Corp., Research Memorandum RM-1726, June 4, 1956.

[17] Baker, R. H., Astronomy, D. Van Nostrand Co., New York, 1950.

[18] The Soviet "Lunik" is an example of such a flight. It missed the Moon with enough velocity to permit it to take up an artificial asteroid orbit.

[19] Industry Invited To Submit Space Capsule Design Proposal, National Aeronautics and Space Administration release, November 7, 1958.

21

Observation Satellites

A. General Application Possibilities

Observation satellites can be useful in military reconnaissance, terrain mapping, astronomical photography, international inspection, cloud observation, and photography of the Earth. Appreciation of the value of such satellites, however, must depend upon some understanding of the kinds of information obtainable, the problems in obtaining information, and the ways of getting this information back from the satellite.

B. Some Historical Perspectives

As soon as man could get a better view from whatever height he had access to, he climbed and used his eyes. When photography became a practical tool—about 100 years ago—he started using cameras from towers, mountaintops, and balloons, and later from rockets and airplanes.

Nadar[1] (1820-1910), a famous French photographer, was a pioneer in aerial photography. In 1858 he started the photographic balloon ascents described in his book, *Les Mémoires du Geánt* (1864). Nadar's views on military applications of balloon reconnaissance changed from a refusal to work for Napoleon III in 1859 to active participation as commander of the balloon corps during the siege of Paris (1870-71).

In 1860 a J. W. Black of Boston joined a "Professor" Sam A. King, a well-known aerialist, to take a balloon photograph of Boston from an altitude of 1,200 feet. This was, for many years, widely regarded as the most successful aerial photograph on record. Oliver Wendell Holmes immortalized this photo with the phrase "Boston as the eagle and the wild goose see it."

General George B. McClellan used balloon photographs in several Civil War battles (1862). He made huge maps, superimposed grids on these maps, and furnished telegraph connection between division headquarters and the balloon-borne observer, anticipating by about 80 years the role of aerial observers for artillery adjustment.

All of the early balloon photographers had rather limited perspectives compared with an American named George Lawrence who started doing aerial photography from balloons in the early 1900's. This remarkable man devised various cameras weighing more than a thousand pounds, taking pictures as large as 4 by 8 feet, and successfully raised them by means of balloons, kites, and associated control apparatus to heights of several thousand feet. One of his earliest cameras was a panoramic camera of the type now proposed for lunar reconnaissance and useful also in observation satellites.[2]

As soon as airplanes were thought practical and safe, photographs were taken from them. In World War I, and more extensively in World War II, photographs were major tools of reconnaissance and intelligence.

Photography from rockets is not a new phototechnique. At a meeting in Stuttgart in 1906, A. Bujard presented a paper, "Rockets in the Service of Photography," describing the work of Alfred Maul, who wanted to use camera-carrying rockets for military reconnaissance.[3] He started with a camera taking pictures 40 millimeters square (the same size as the picture taken by a miniature Rolliflex camera). In spite of many difficulties he devised in 1912 a rocket stabilized by a gyroscope, with a takeoff weight of 92.5 pounds. This rocket carried an 8- by 10-inch camera to about 2,600 feet. By this time, however, the airplane was coming into its own, and photos from airplanes were easily made. The success of airplanes caused loss of interest in rockets as photographic platforms. We are now witnessing, in current interest in observation satellites, a return swing of the pendulum.

C. Intelligence and Reconnaissance

In satisfying both national and military intelligence requirements, data are needed on a real or possible enemy's capabilities and intentions. Included in the data is information that will help answer such questions as: What does the enemy have? Where is it? How many does he have? How does he use it? How good is he at using it? Is he about to use it? etc. Further needs are for mapping, charting, and weather reconnaissance.

Data which ultimately can be used to produce intelligence can and do come from many sources of varied character.[4] Reconnaissance is one possible source. Spaceborne reconnaissance will be valuable because of the extremely large areas that can be covered rapidly, and the possibility of cyclic operation.

D. Sensors for Use in Observation Satellites

Observation can be carried out in many ways, using sensors operating in different portions of the electromagnetic spectrum (see section "Scientific Space Exploration"). The first sensor used by man was the unaided eye. Next came photography with its ability to record in permanent form large amounts of detailed information. Then followed the development of radar reconnaissance, electronic interception, and infrared reconnaissance.[5]

For reconnaissance in peacetime the all-weather requirements normally associated with combat operations can usually be waived. Because ground objects of military interest are, in some important cases, growing smaller (e.g., missile sites compared with airfields), there is a growing requirement for very refined detail in reconnaissance. These considerations point to photography as the most likely tool for satellite operations.

E. Ability To See from an Observation Satellite

Various factors enter into an estimate of the degree of detail that can be detected or identified[6] by a visual sensor system.

Among these are distance from the sensor to the object viewed and the focal length of the viewing lens, from which is usually computed a *scale number*. The scale number, S, is the ratio of altitude to focal length:

$$S = \frac{\text{Altitude}}{\text{focal length}}$$

For example, consider a camera with a focal length of 6 inches at an altitude of 150 miles. The scale number of the photograph is then:

$$S = \frac{(150 \times 5{,}280)\ \text{feet}}{0.5\ \text{foot}} = 1{,}584{,}000$$

This means that 1 inch on the photograph corresponds to S inches on the ground, or about 25 miles. In general, the larger the scale number, the harder it is to see fine detail. For pictures taken other than straight down, scale numbers vary from point to point, getting larger as the view moves toward the horizon.[7, 8] Figure 1 and table 1 should give an appreciation of the distances, viewing angles, and ground coverage possible from some selected altitudes.

Another useful parameter is *resolution,* a term originally used by astronomers to specify the ability of a telescope to separate double stars. As applied to photographs, resolution refers to the ability of a film-lens combination to render barely distinguishable a standard pattern consisting of black

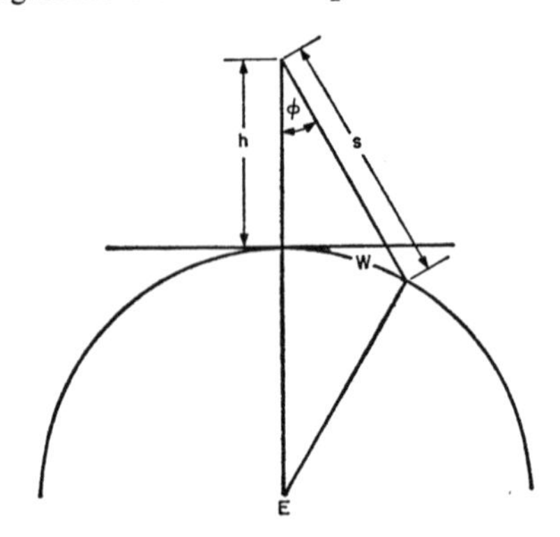

FIG. 1. Geometry of coverage from a satellite

and white lines. When a lens-film combination is said to yield a resolution of 10 lines per millimeter, it means that it can make distinguishable lines and spaces when there are 10 of them per millimeter. Lines of coarser spacing are seen more clearly. There are limitations on the usefulness of this single parameter for analytical purposes, but it is a convenient basis for gross comparison of sensing systems.[9-18]

TABLE 1.—*Distances, viewing angles, and ground coverage for satellite altitudes*

Satellite altitude, h (miles)	Viewing angle, ϕ (degrees)	Slant range, s (miles)	Ground distance, W (miles)
100	30	116	58
	45	143	101
	60	208	180
200	30	233	116
	45	290	205
	60	436	378
300	30	351	176
	45	442	312
	60	690	599
500	30	590	296
	45	758	538
	60	1,349	1,186
1,000	30	1,208	607
	45	1,667	1,198
	53	3,021	2,558
2,000	41.6	4,496	3,342
5,000	26.2	8,093	4,410
10,000	16.5	13,448	5,075

NOTE.—The last entries in the table for 1,000 miles and greater altitudes are the maximum values of ϕ, s, and W attainable from that altitude.

Ground resolution, G, is often used in discussing performance. It is simply the ground-size equivalent to one line, at the limit of resolution. Thus, if a film-lens combination yields R lines per millimeter, which means a line, at the limit of resolution, is $1/R$ (millimeters) and the scale number is S, the ground resolution is S/R. To use familiar units, and rounding off a bit—

$$\text{Ground resolution (feet)} = \frac{S}{300\,R\text{ (lines per millimeter)}}$$

For the same focal length and altitude used in the example above, if the film resolution is 100 lines per millimeter, the ground resolution is about—

$$G = \frac{1{,}500{,}000}{300 \times 100} = 50 \text{ feet}$$

From the formula for ground resolution one would expect to obtain the same ground resolution by trading resolution and scale number. Thus one should expect that 10 lines per millimeter at a scale number of 100,000 should yield the same ground resolution as 100 lines per millimeter at a scale number of 1 million. However, this type of reciprocity is never the case, either in practice or in theory—if one can trade scale for resolution in a design, he should trade in the direction of lower resolution and smaller scale number. There are great differences in the graininess characteristics of different aerial photographic emulsions, and these affect interpretability much more than they influence resolution.

It is convenient to define four levels of photographic detail: A, B, C, and D. These levels are, in terms of ground resolution:

A: 50 to 200 feet.
B: 10 to 40 feet.
C: 2 to 8 feet.
D: 0.5 to 2 feet.

The range of resolution within each level arises from a practical inability to measure and interpret ground resolution as a fixed number, and from additional detailed factors, such as graininess of photographic emulsions.

Photographic operations at resolution level A would be useful in covering large areas, measured in millions of square miles. From such pictures one should be able to see and identify most lines of communications, railroads, highways, canals, urban centers, industrial areas, airfields, naval facilities, seaport areas, and the like.

Level B would be appropriate to covering areas measured in hundreds of thousands of square miles. With such resolution, the character of many major installations could be detected and identified; aircraft could be seen on airfields; almost all lines of communication could be found and plotted; and, in general, items merely detected at level A could be seen more satisfactorily. It would be desirable, of course, to eliminate category A and cover millions of square miles at resolution level B.

Level C is indicated for areas of specific interest measured in terms of hundreds of square miles. At this level, extremely detailed analyses of sites, airfields, industries, and activities could be made. Most World War II photography was accomplished at this level of detail.

Level D is appropriate for rather small areas, say, of the order of 1 square mile. Such operation would provide data in very fine detail about new activities, sites, and installations.

A reasonable concept of observation satellite operation would cover all areas of interest at level A (or, preferably, level B) at intervals of perhaps 6 months to a year. With such an operation, new major installations could be detected, perhaps patterns of use found, and hints and clues obtained for the direction of other, higher resolution reconnaissance systems. An overall observation capability, whether for reconnaissance or inspection, should be based upon a family of systems able to operate at each of the levels indicated.

It is extremely difficult, if not impossible, to take a given number—such as a ground resolution of, say, 80 feet—and describe specifically what can be seen. The conditions of observation—the illumination, the contrast, the context, and many other important factors that determine detection and identification, are so variable that specification of ground resolution alone is insufficient. For example, rocket photographs are available that were taken with a 6-inch lens from an altitude of about 150 miles, with resolution estimated to be about 10 lines per millimeter—indicating a ground resolution of about 500 feet.[19] This is much poorer than level A. On one of these photographs (figure 2), however, major railroads show up clearly (figure 3) to even casual observation; two major airfields are easily seen, with runways clearly distinguishable; and major streets in a nearby city are fairly easily resolved and can be plotted. (This is an example of the phenomenon of long lines being more easily detected than small square objects.) The only thing that is really clear is that small ground resolution is better than large ground resolution—

much more detail can be seen with a system yielding 10-foot ground resolution than one yielding but 50-foot ground resolution.

It is realistic to suppose that one can find large installations in the process of construction by using systems performing at level A or B. At such levels it is certainly possible to get clues of sufficient interest to warrant dispatching a system operating at levels C and D to verify, confirm, or further inspect these operations.

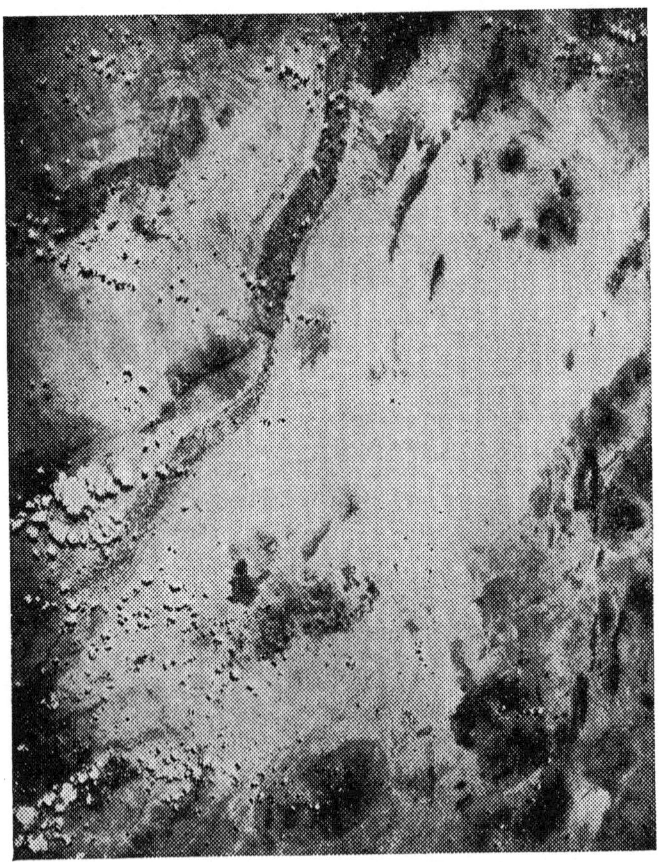

FIG. 2. Rocket photograph

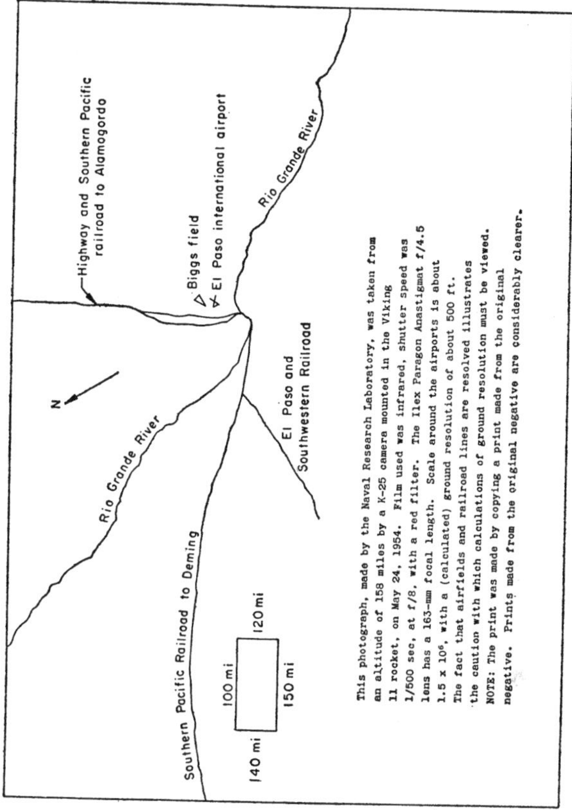

FIG. 3. Analysis of photograph (Fig. 2)

Observation at level A is a fundamental and first-thing-first job. It could supply a matrix in which other data can be imbedded, and it could furnish a planning guide for future reconnaissance.

F. Cameras Suitable for Satellites

The camera used to make the photograph of figure 2 is now considered obsolete. Lens designs, camera precision, and film characteristics have all improved since the time of its construction. New types of cameras, in particular the shutterless, continuous-strip camera and panoramic cameras, have emerged as important and useful tools.[20-22] Modern panoramic cameras are derivatives of a rather old camera design, wherein a lens, which itself covers only a narrow angle, sweeps out a wide-angle photograph by sequentially scanning the scene and depositing the resultant image on a relatively long but narrow strip of film. A basic characteristic of strip or panoramic cameras is the fact that the photograph is taken sequentially, and possible distortions thus produced make precise measurements extremely difficult. Cameras with between-lens shutters avoid this difficulty, and are universally used for mapping.[23]

Increased speed of modern aircraft has motivated development of methods for compensating for image speed during exposure.[24] Image speeds, for cameras mounted vertically in a satellite, may be readily calculated. At an altitude of 300 miles, for example, satellite velocity is about 25,000 feet per second. The image speed is simply the vehicle speed divided by the scale number. As an example, consider a 6-inch focal-length camera at 300 miles altitude. The scale number is about 3 million. The image speed is therefore about 0.1 inch per second.

To achieve high resolution—say 100 lines per millimeter—the blur caused by uncompensated image motion must be restricted to less than about one two-hundredths millimeter.[25] This can be accomplished by a combination of image-motion compensation and short exposure time.

If sensitive film emulsion or extremely high-speed optics are used, exposure times in satellite observation of the order of one five-thousandths second are not unreasonable. Clearly, with a very short exposure time, blur is minimized. Even the high satellite speed of 25,000 feet per second will yield a blur corresponding to only about 2½ feet on the ground, if exposure time is one five-thousandths second.

There has also been progress in lens design. Camera lenses with focal lengths of 100 and 240 inches have been developed by the Air Force.[26] Such lenses can be expected to find eventual utility in satellite observation systems to obtain very fine ground resolution.

An idea of what can be seen with comparatively small ground resolution can be gained from figure 4, with a calculated ground resolution of 48 feet. This photograph was taken from a balloon at an altitude of 87,000 feet. The camera used film 70 millimeters wide, and the lens had a focal length of about 1.5 inches. The scale number is 700,000. This photograph covers an area approximately 24.9 miles on a side (620 square miles). Objects such as dam sites, railroads, farm ponds, and highways are easily distinguishable, as are some of the larger structures in the town of Truth or Consequences, New Mexico. Agricultural patterns too are readily distinguishable. This experiment in high-altitude photography is described in the reference noted below.[27]

The following table of approximate resolutions and focal lengths for a satellite at about 150 miles altitude is of interest, especially when comparing the calculated numbers with those applicable to figure 4 (ground resolution of 48 feet).

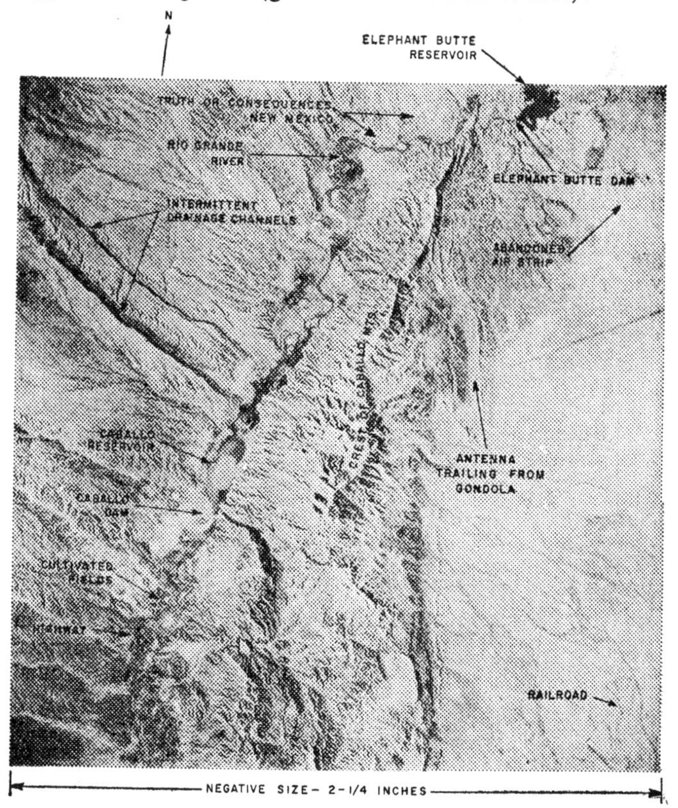

FIG. 4. Balloon photograph

TABLE 2.—*Approximate resolutions and focal lengths, satellite at 150-mile altitude*

Focal length (inches)	Scale number	Ground resolution (feet)— At 40 lines per millimeter	Ground resolution (feet)— At 100 lines per millimeter
12	750,000	60	24.0
36	250,000	20	8.0
120	75,000	6	2.4

Comparison of this table with the actual photograph of figure 4 is striking evidence of the ability of satellites to secure useful photography.

Mapping photography from 30,000 feet during World War II secured a ground resolution of perhaps 15 to 20 feet. Thus, satellite performance possibilities are very attractive in these terms also.

Throughout this discussion, reference has been to photographic film systems for consistency of illustration only. Television techniques are also entirely applicable to use in observation satellites, and the concepts, formulas, and calculations presented apply also to television systems.

Photographic film exposed in a camera aboard a satellite, like film exposed in any camera, would have to be processed before the data can be used. Conventional automatic processing techniques could be used, with the film being stored until the satellite is in the vicinity of a ground receiving station. It could then be scanned, using television techniques, and the data transmitted to Earth. Received on Earth, it could be stored as a video tape, or reconstituted into a photographic record for interpretation, study, and analysis at a suitable time and place.

If a television camera system were used, a similar problem would arise. The data would have to be stored—on magnetic tape, for example—to be read out and transmitted to a ground station at an appropriate time.

Photographic film could also be returned to Earth directly for processing. The use of television techniques in the case of physical recovery of data is not clear.

G. Returning Data from Observation Satellites

There are two distinct ways in which information picked up by a sensor in a satellite can be returned to Earth: it can be transmitted by radio or physically returned to Earth in a data capsule.

Choice of the method of return depends upon such factors as the requirements for timeliness, the rate of information flow, the total volume of information, and the form in which the information is to be used. There certainly may be military reconnaissance or international inspection operations which require minimum time delay between observation and transmission of data. An obvious example is a satellite system to observe missile firings as part of an international system for warning of surprise attack. The useful life of such information is short—a matter of very few minutes—and direct radio transmission of data is rather clearly indicated.

Inspection of large areas, or detailed examination of smaller areas, may result in the collection of volumes of data so large as to saturate the radio transmission capability of a satellite. Physical recovery might well be a better way to deliver such data. Mapping, where the principal concern is with geometric fidelity, is another case where physical data recovery may be better than radio transmission.

Preliminary to an attempt to assess the relative merits of either of these recovery methods, it is necessary to reconcile the many ways of estimating the information content of a photograph. From the viewpoint of the communications engineer, a photograph represents a calculable number of "bits" of information; this number permits calculation of the capacity of the communication channel and the time required to transmit the photograph.

The photointerpreter tends to evaluate the photograph in terms of the useful detail he can find and the extent of the coverage. The aerial photographic scientist will think in terms of scale, resolution, and area covered.[28]

The relationships between these viewpoints permit some valuable comparisons to be made. It is specifically interesting to consider the way in which one can compare a photograph and television image for the purpose at hand. Commercial high-quality television transmission systems use a bandwidth of about 4.25 megacycles per second. A high-quality photograph will have a resolution of about 100 lines per millimeter. Information theory shows that—

> A communication system with a bandwidth of 4.25 megacycles per second will have to operate continuously for 32 minutes to transmit the quantity of information that can be stored on a single 9- by 9-inch photograph at 100 lines per millimeter.

Satellite systems can readily be postulated that would collect in 1 day many times more data than could be transmitted by video link. For example, consider a system with a 36-inch focal length (this is a median example: for mapping, much shorter lenses could be used; for extremely detailed observation, much longer lenses would be needed) which covers (either through single lens panoramic techniques or through a conventional multiple camera installation) an angle of about 90° from about 150 miles altitude. Such a system could cover about 3 million square miles per day, and consume at least 1,500 feet of 9-inch wide film. If a resolution level of 100 lines per millimeter is obtained, this amount of information would require over 1 month to transmit, operating 24 hours per day. Such a photographic system seems to be well within the state of the art. Systems with larger capacity are conceivable and may prove desirable.

Clearly there are important cases where a video link simply cannot transmit information as fast as it can be acquired. Small improvements in transmission systems, increasing the numbers of ground stations, etc., can make only minor improvements in a serious and fundamental limitation. This is one of the main reasons for the importance of physical recovery of a film payload.

For the collection of great quantities of data over short periods of time, it would be reasonable to use a photographic satellite whose data could be physically recovered after a short orbital life. Satellites which transmit data by video link are better suited for long-lived surveillance operations. The two types of satellite are complementary, not competitive; together they would constitute a balanced satellite reconnaissance capability.

H. Mapping from Satellites: Some Preliminary Considerations

The problem of mapping is radically different from most other kinds of observation. Ground resolution, however calculated, is an overconservative statistic by which to measure the performance of a mapping system. To map a surface with geometric accuracy, "hard photography" free of distortions is needed. Mapmakers have generally required wide-angle photography with a highly corrected, essentially distortionless lens, the entire photograph being taken simultaneously (that is, with a between-lens shutter) instead of sequentially as with focal-plane shutters, strip photography, and panoramic photography.[29] Distortionless wide-angle optics, capable of yielding 50 lines per millimeter on film over the entire field, are now available.

While the requirements of mapping are very stringent with respect to distortion, they are rather modest as regards ground resolution. The distance between 2 points may be measured on a photographic plate to within an error perhaps 10 times less than the resolution figure. For example, a system with a ground resolution of 2,500 feet could yield a mapping precision of about 250 feet. Such resolution can be obtained, for example, from an altitude of 4,000 miles with focal length of 6 inches, using 50-lines-per-millimeter photography.

There are many other problems associated with mapping. In particular, the use of film instead of glass plates (as would be necessary in satellites) imposes the requirement for careful registration marks in the form of a *réseau* or fine grid on the plate, carefully calculated and fiducial marks along the edges of the format. It is felt that techniques such as these can yield precision measures. Measurements to 2 seconds of arc should be achievable.

These numbers can be used to go up and down the scales of altitudes and focal lengths. For example, coming in to an altitude of 1,000 miles with a 6-inch lens gives numbers 4 times better than those given above. Coming in to 1,000 miles with a 12-inch lens (not unreasonable for a mapping satellite) gives numbers 2 times better than these. Thus, in principle, we could measure to perhaps 30 feet. The speed in orbit of a satellite at 300-mile altitude is about 25,000 feet per second.

An exposure time of one five-hundredth second implies a maximum theoretical blur corresponding to about 50 feet on the ground, which yields an effective blur of about 25 feet. At much greater altitudes, the orbital speed is even less. This amount of blur could be ignored or, alternatively, a very simple and fixed amount of image motion compensation could be applied to the film during the exposure to eliminate this blur.

The flat-Earth approximation would yield a ground coverage for the conventional mapping angle of 76° fore and aft of 1½ times the altitude; since the Earth is curved, the coverage is in excess of this. It would not take very many pictures from altitudes of 1,000 miles or more to map the Earth successfully.

It is important to recognize that we can measure distances with accuracy greater, in principle, than that specified by the resolution limit as calculated above. Furthermore, excellent modern wide-angle optics are available.

A mapping satellite would likely require a platform stabilized with respect to the horizons. It would require recovery. It is mandatory to return a precision photograph directly for examination rather than to incur possible geometric degradation via an electronic relay station. Auxiliary ground tracking stations, and perhaps visual and electronic beacons aboard the satellite, might be required and useful to locate the satellite precisely, as might simultaneous star photography for precision determination of the satellite attitude at moments of exposure.

In general, the fewer photographs which have to be tied together to map a given area the better. Thus, with mapping lenses capable of photographing approximately a 1,500- by 1,500-mile square from 1,000 miles altitude, the number of photographs required to map even huge areas is small. If the photographic sequence starts in an area which is mapped, the map can be extended to the new areas.

I. Photographic Aspects of Meteorological Observations from Satellites

Progress in satellites for intelligence or inspection purposes will inevitably proceed in the direction of higher resolution systems selectively covering smaller regions on the ground. Weather reconnaissance, on the other hand, can be done at resolutions measured not in feet but perhaps in hundreds of feet. Instead of narrow-angle views of the Earth's surface, it requires extremely wide-angle views, covering as much as possible simultaneously. These rather different requirements indicate that the purpose of both fields would be served by separate satellites rather than by trying to develop one all-purpose type.

J. Observation Satellites and Inspection

Proposals for utilizing aerial inspection systems as a part of atomic-energy controls were made as early as 1947 in a United Nations report.[30] Since that time, aerial inspection systems, used alone or in conjunction with ground systems, have been proposed for disarmament inspection and surprise-attack warning systems.

Observation satellites could certainly contribute to, if not, in fact, provide, an inspection system.[31-35] An observation satellite could be used to detect and locate missile firings, for example, serving either to monitor agreements regarding missile launchings or to assist in reducing the measure of surprise from an attack. In conjunction with aircraft, satellites could perform installation inventory, assist ground inspectors in locating previously unknown or new military sites, monitor shipping, and carry out numerous other tasks.

K. Some Possible Observation Satellite Combinations

Attempting to list all possible kinds of observation satellites—taking note of varying operational altitudes, sensors, purposes, methods of returning data, etc., makes for a bewildering array. Table 3 represents an attempt to sort out and classify some of the possible combinations. It is in no way intended as a complete listing.

TABLE 3.—*Some possible observation satellite combinations: a rough outline*

Basic booster component	Typical weight on orbit (pounds)	Typical altitude (miles)	Typical sensor	Typical ground resolution[1] (feet)	Useful life: Short (week or less)	Useful life: Long (month or more)	Data recovery: Physical recovery	Data recovery: Video link	Purpose
IRBM	300-500	150	Photo	60	X		X		Coverage of millions of square miles (level A).
		150	Photo	20	X		X		Higher resolution search over limited areas (level B).
		300	Photo-TV	200-500		X		X	Weather reconnaissance.
ICBM	2,000-10,000	300	Photo-TV	200-500		X		X	Weather reconnaissance.
		150	Photo	20	X		X		Higher resolution coverage of millions of miles (level B).
		300	Photo-TV	8-12		X		X	Cyclic surveillance of selected areas—warning (?) (level C).
		300	Infrared			X		X	Warning (ICBM firings).
		300	Electronic			X		X	Electronic and communication intelligence, communication relay, etc.
Nuclear or large chemical rocket	20,000-100,000	500-25,000	All types	(2)		X	X	X	All missions listed above plus level D.

1 Ground resolution figures apply only to photographic and television sensors.

2 1 foot at 500 miles.

L. Scientific and Civil Uses of Observation Satellites

Observation satellites can serve uses other than military reconnaissance, inspection, weather forecasting, and mapping. They can also make observations of other celestial objects far superior to those obtainable within the Earth's atmosphere.[36]

The advantage to astronomical photography of reducing atmospheric effects has been shown dramatically by Sun photographs from a balloon-borne telescope.[37] Project Stratoscope employed a special photographic telescope weighing 300 pounds, and photographed the Sun from 80,000 feet, above about 90 percent of the Earth's atmosphere. According to the project scientists, the photographs taken of the Sun are the sharpest ever secured, showing Sun detail never before seen.

The enormous advantages of clear photographs of celestial objects can be appreciated from a comparison of the pictures of figure 5. The photograph of the Moon shown in this figure shows the region of Clavius. Taken by the 200-inch telescope on Mount Palomar, it is considered to be one of the finest photographs ever made. The ground resolution is about 1 mile. In an effort to show how little lunar detail we can readily see in such photographs, despite their high quality, a photograph of Washington, D. C., was systematically degraded to show ground resolutions of 200, 500 and 1,000 feet.[38] Note the low level of detail of Washington at the 1,000-foot resolution level. This is five times better than the "sharp" Moon photograph.

Aerial photography of the Earth has been applied to exploration; Earth sciences; land planning; crop, soil, and forest inventories; engineering; ecology; geology; geography; physiography; geomorphology; hydrography; urban-area analysis and planning; and archeology.[39-44]

The application of aerial photography to these varied fields depends first upon the large view afforded and on the recording of fine enough detail to permit accurate identification, measurement, and comparison.

Satellites will yield a grander view, a larger perspective, than has ever been attained before. Photographs from rockets at altitudes of 150 miles have already yielded spectacular views. The possibility of seeing, as a whole, relationships, for-

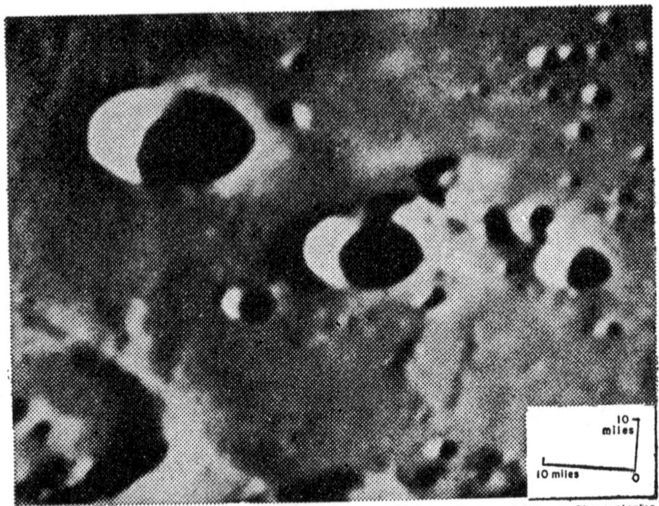

200-inch photograph, Mount Wilson and Palomar Observatories

The Floor of Clavius

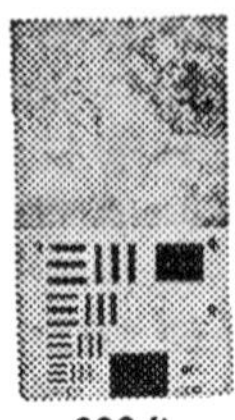

200 ft
Ground resolution

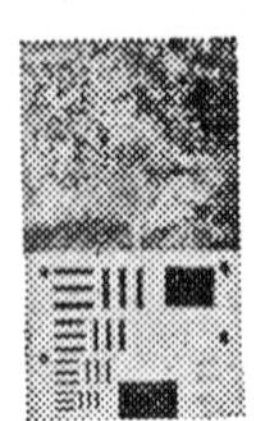

500 ft
Ground resolution

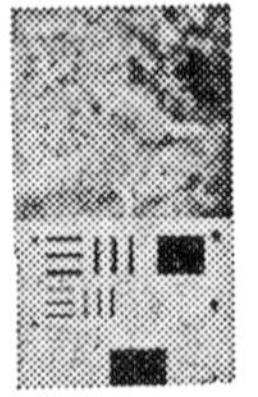

1000 ft
Ground resolution

0 10 miles

Washington, D. C.

FIG. 5. Lunar photography and aerial photography: resolution comparison

mations, and terrain features which require the perspective of distance is an exciting prospect. The world today is still poorly mapped, and its resources still not measured.

At least several novel applications of observation satellites can be foreseen. Ice and snow surveys over vast areas, iceberg patrol, and studies of ocean-wave propagation—all require the coverage of large areas hitherto impossible by conventional airborne observation systems.

The emergence and utilization of such a radically new tool as the observation satellite will undoubtedly result in the development of applications and techniques not yet imagined or foreseeable.

M. The 24-Hour Satellite

The satellite whose period is exactly that of the Earth—commonly called the 24-hour satellite—is of special interest. If placed over the Equator, moving eastward in its orbit, it would appear to remain always at the same point in space, as viewed from the Earth. The orbit altitude required for this 24-hour period is about 22,000 miles. Such a satellite would have in view, at usable angles of incidence, about 38.2 percent of the Earth's surface, or approximately 75 million square miles (figure 6).

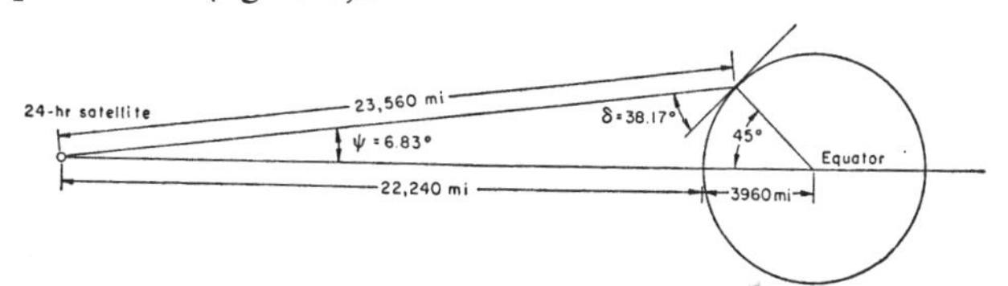

FIG. 6. Viewing an area on the Earth at 45° latitude from the "24 hour" satellite

Interest in the 24-hour satellite for observation stems mainly from the somewhat intuitive notion that it would be useful to observe activities on Earth from a fixed position. However, it is not altogether clear that useful observations can be made from such a great distance. A worthwhile preliminary examination of the problems involved can be conducted by assuming a desire to achieve observations with a ground resolution of 100 feet. This is a better order of resolution, considering the viewing distance, than is commonly achieved at astronomical observatories such as Mount Wilson and Mount Palomar.

However, the troubles that limit achievable resolution at Earthbound observatories are, in large part, chargeable to the atmosphere. The optical system in a 24-hour satellite will have many problems, but not this one. The problem here is simply one of size. Taking into account practical limitations on photographic film, etc., it appears that achievement of the assumed resolution of 100 feet would require a lens diameter approaching 8 feet, with a focal length of about 64 feet. Such optics are huge, even by terrestrial standards. This is the size of the Mount Wilson telescope.

Suppose that such a camera were used to take photographs 4 by 5 inches in size of an area at about 45° latitude. The region photographed would be about 152 by 197 miles, an area of approximately 30,000 square miles. This is only slightly less than the combined area of Connecticut, Massachusetts, New Hampshire, and Vermont.

Clearly, the achievement of ground resolution such as the one calculated above would be an engineering and scientific triumph of the first magnitude. The requirements for stability of the camera system, given any reasonable exposure time (such as one-hundredth second), are severe. Angular motions which tend to blur the photograph must be kept to, say, 0.05 second of arc during the exposure time. This is a formidable requirement, indeed.

There is one application of a 24-hour satellite that would not require such huge optical systems: weather observation. Here, relatively poor ground resolution would be adequate; furthermore, an enormous view would be afforded from such a high altitude.

N. Uses and Requirements of Very Large Observation Satellites

A need for very large satellite payloads—say, of the order of 100,000 pounds—can already be discerned. Current thinking about payloads for observation satellites has been shaped to a considerable extent, not by the inspection of observational needs, not by optics design, but by booster availability and performance. Thus, designers of observation satellites have first looked at rocket vehicles in development, and attempted to use whatever weight such vehicles may allow them to put on orbit.

These weights have, to date, been fairly small. Roughly,

they range from something like 20 pounds to a few tons.

Payload limitations constrain the performance of an observation satellite in three respects: quality of data obtainable; quantity of data secured; and timeliness with which these data can be delivered to the user. Current payload limitations preclude observations from satellites of the sort available from aircraft. Ground objects of interest have been observed successfully with reconnaissance equipments operating in aircraft at altitudes measured in miles, not hundreds of miles. The size of the photographic gear useful at high aircraft altitudes (say, of the order of 50,000 feet) are very large, with weights from several hundred pounds to several thousand pounds. As the camera is moved away from the observed objects to distances 15 to 100 times aircraft altitudes, the size of equipment needed for similar observation quality is going to grow sharply—indicating satellite payloads of many tons. The need for satellite payloads of this order is supportable by the argument that observational capabilities of the kind now available only to aircraft will continue to be needed, perhaps with growing urgency. Ground objects likely to be of inspection or observational interest in the future will probably not get larger than objects of current interest and may, in fact, get smaller.

Figure 7 shows the relationship among altitude, focal length, and ground resolution on the assumption of a film resolution of 100 lines per millimeter. If poorer resolution is obtained on the film, the required focal lengths increase proportionately.

Observations under twilight, moonlight, and eventually under night-sky illumination only, are goals no longer considered unattainable. There are areas of the world that are continuously dark for considerable periods. Observation, inspection, and surveillance of these areas from satellites might prove impossible unless needed improvements in observation technology are pursued. Nighttime satellite observation would be desirable, of course, to counter attempts to employ darkness as a shield for various activities. Promising developments, such as certain refined television techniques, point to one way of achieving such goals. Photography from airplanes by the light of a near-full Moon has been accomplished and described by the Air Force. Either of these kinds of approaches would involve large-diameter, long-focal-length op-

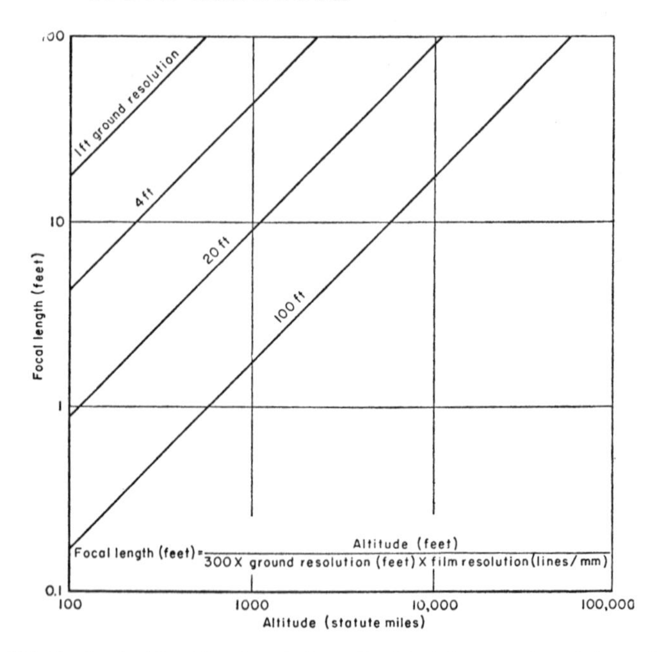

FIG. 7. Required focal-length variation with altitude for ground resolutions of 1, 4, 20, and 100 feet assuming that 100 lines per millimeter are obtained on the film

(Note that the speed of the optical system—the f/ number—is left out. The high resolution used above requires optics of at least f/8, preferably faster)

tics to collect as much light as possible and recoup quality loss inherent in low-illumination operation. A speculative example might be a camera system with 100-inch focal length, at a speed of f/2, thus requiring optics at least 50 inches in diameter. In addition to auxiliaries such as power supply, controls, and communication equipment, any large, fast optical systems such as this one would require automatic focusing mechanisms, temperature control, and other refinements not needed for smaller, slower, less sensitive systems—all of which points to payloads of many tons.

Better observational quality—finer detail—will require weight. Greater total quantity of information also requires weight. Information quantity depends on total vehicle life, the total load of film or other storage media carried, and the method of returning data to earth. Vehicle useful life depends, among other things, upon the life of its power supply. For those satellites powered by chemical batteries alone, the upper limit of working life is set by the total energy content of the batteries and the rate at which such energy is used by the apparatus in the satellite. Working life would increase if more batteries (weight) were added. When such power supplies can be supplemented by solar energy, or when the power requirements can be cut down, working life will increase for a given total weight. Nuclear powerplants will eventually be available, and will save weight. They may also create added problems, calling for added weight for shielding.

The "quantity" of information deliverable by a satellite which communicates with the ground by video link is directly dependent on the bandwidth used and the time available for communication. Generally speaking, greater bandwidth requires more equipment, hence more weight. The available time for communication is determined by the number of ground stations, and the fraction of time the satellite is within view of a given ground station. For 300-mile-altitude satellites, this time is about 10 minutes per station. Higher altitude satellites are in view longer, and in addition have less orbital speed. Thus a 1,000-mile-altitude satellite would be in view for more than twice the time of a 300-mile satellite. Very high satellite altitudes may also permit direct line-of-sight communication from satellite to ground stations at the moment that the satellite is picking up data. This might indeed eliminate the need for data storage aboard the satellite, and would certainly be desirable for warning applications. Higher orbit altitudes require large launching rockets—the same eventual need as that manifested by payload increases.

Environmental factors, about which all too little is known at present, include the radiation environment at considerable altitudes. The early data available from Explorer IV are fragmentary, and much more information is needed. It may be suspected that the operation of a film-carrying satellite in the indicated heavy radiation field will prove to be extremely difficult. The implication of the data at hand is that considerable shielding will be needed. Shielding is directly translatable into pounds of payload. If photographic film proves to be a poor sensor to use under such circumstances, a change to another sensor such as television and electrostatic tape may be required. These would undoubtedly yield poorer resolution than photographic film; the loss could be recouped by using larger, heavier equipment.

If a man is to be put aboard an observation satellite, the total weight required will climb sharply.

Multiple sensors (i.e., combinations of photographic, infra-red, radar, and other sensors) in satellites would be of value, and would certainly require larger vehicles than those needed for each of these separately.

One can envision a need for measures to decrease the vulnerability of satellites by changing their orbits, which might require re-aiming the cameras. All of these would require weight, hence larger launching rockets.

In brief, large launching rockets are an inevitable requirement if observation satellites are to progress vigorously.

Notes

[1] Pseudonym for Gaspard Felix Tournachon.

[2] Davies, M. E., A Photographic System for Close-up Lunar Exploration, The RAND Corp., Research Memorandum RM-2183, May 23, 1958.

[3] Ley, W., Rockets, Missiles, and Space Travel, revised edition, The Viking Press, New York, 1957.

[4] Katz, A. H., Contributions to the Theory and Mechanics of Photo-Interpretation from Vertical and Oblique Photographs, Photogrammetric Engineering, vol. XVI, No. 3, June 1950, pp. 339-386.

[5] See footnote 4.

[6] A common view held by experienced workers in aerial photography is that detection is much "easier" than identification. For example, one might be able to detect objects on a road, and be unable to decide whether they are trucks, tanks, or smaller vehicles. It usually requires roughly five times better ground resolution to identify objects.

[7] See footnote 4.

[8] Katz, A. H., The Calculus of Scale, Photogrammetric Engineering, vol. XVIII, No. 1, March 1952, pp. 63-78.

[9] See footnotes 2, 3, 4, 6, and 8.

[10] Brock, G. C., Physical Aspects of Air Photography, Longmans, Green & Co., New York, 1952.

[11] Schade, Otto, Electro-Optical Characteristics of Television Systems, RCA Review, vol. IX, Nos. 1-4, March-December 1948.

[12] Higgins, G. C., and R. N. Wolfe, The Relation of Definition to Sharpness and Resolving Power in a Photographic System, Journal of the Optical Society of America, vol. XLV, No. 2, February 1955.

[13] Higgins, G. C., and R. N. Wolfe, The Role of Resolving Power and Acutance in Photographic Definition, Journal of the Society of Motion Picture and Television Engineers, vol. LXV, No. 1, January 1956, pp. 26-30.

[14] MacDonald, D., Some Considerations of Resolution, Sharpness, and Picture Quality in Technical Photography, Photographic Society of America Journal, vol. XIX, No. 5, May 1953, pp. 49-55.

[15] MacDonald, D., Air Photography, Journal of the Optical Society of America, vol. XLIII, No. 4, April 1953, pp. 290-298.

[16] Optical Image Evaluation, proceedings of the NBS Semicentennial Symposium on Optical Image Evaluation held at the NBS in October 1951, National Bureau of Standards Circular 526, April 1954.

[17] Rose, A., Television Pickup Tubes and the Problem of Vision, Advances in Electronics, vol. I, 1948, pp. 131-165.

[18] Katz, A. H., Air Force Photography, Photogrammetric Engineering, vol. XIV, No. 4, December 1948, pp. 584-590.

[19] Baumann, R. C., and L. Wunkler, Rocket Research Report XVIII—Photography from the Viking XI Rocket, NRL Report 4489, Naval Research Laboratory, February 1, 1955.

[20] See footnote 18.

[21] Katz, A. H., Aerial Photographic Equipment and Applications to Reconnaissance, Journal of the Optical Society of America, vol. XXXVIII, No. 1, July 1948, pp. 604-610.

[22] Goddard, G. W., New Developments for Aerial Reconnaissance, Photogrammetric Engineering, vol. XV, No. 1, March 1949, pp. 51-72.

[23] Katz, A. H., Camera Shutters, Journal of the Optical Society of America, vol. XXXIV, No. 1, January 1949, pp. 1-22.

[24] See footnotes 10 and 18.

[25] See footnote 23.

[26] See footnote 22.

[27] DiPentima, A. F., High Altitude Small-Scale Aerial Photography, Rome Air Development Center, Technical Note 58-165, ASTIA Document No. AD-148777, July 1958.

[28] See footnote 4.

[29] See footnotes 4, 16, and 23.

[30] Second Report to the Security Council, U. N., AEC, September 11, 1947 (AEC-36, September 11, 1947), pt. II.

[31] Leghorn, Col. R. S., No Need To Bomb Cities To Win War, U. S. News & World Report, January 28, 1955, pp. 78-94.

[32] Leghorn, Col. R. S., U. S. Can Photograph Russia from the Air Now, U. S. News & World Report, August 5, 1955, pp. 70-75.
[33] Final Report, Subcommittee on Disarmament, Committee on Foreign Relations, U. S. Senate, 85th Cong., September 3, 1958, p. 15.
[34] The Congressional Record, vol. 104, No. 18, February 4, 1958, p. 1389.
[35] The Congressional Record, vol. 104, No. 18, February 4, 1958, pp. 1401-1402.
[36] Introduction to Outer Space, the White House, March 26, 1958.
[37] Project Stratoscope Reveals Sun Data, Aviation Week, October 28, 1957.
[38] See footnote 2.
[39] Gwyer, J. A., and V. G. Waldron, Photointerpretation Techniques—A Bibliography, Library of Congress, Technical Information Division, March 1956.
[40] Selected Papers on Photography and Photointerpretation, Committee on Geophysics and Geography, Research and Development Board, Washington, D. C., April 1953.
[41] Report of and Proceedings Commission VII—Photographic Interpretation to the International Society of Photogrammetry, Washington, D. C., September 1952.
[42] Manual of Photogrammetry, 2d ed., The American Society of Photogrammetry, Washington, D. C., 1952.
[43] Gutkind, E. A., Our World from the Air, Readers Union, London, 1953.
[44] Le Lause, P. C., La Decourverte Aérienne du Monde, Horizons de France, Paris, 1948.

22

Meteorological Satellites

Weather affects almost every known activity of man either directly or indirectly; and the probability of success of many enterprises, civil and military, can be noticeably increased if the weather factor can be counted as a "known parameter."

The various weather services are constantly striving to provide this knowledge, either in the form of actual data concerning current weather or in the form of a forecast of future conditions. In either case, the information provided is only as good as the weather data available to the meteorologist.

It is well known that there is in operation a worldwide weather data collection network supported by most of the civilized nations of the world. Furthermore, these data are freely disseminated to all nations participating in the collection program. While there appears to be a great wealth of information available, it is unfortunately true that there are large areas of the world (e.g., the oceans and the polar regions) from which very little day-to-day weather information is forthcoming. The only means presently available for filling these gaps in weather data is reconnaissance by aircraft or the positioning of weather observing ships. At best, even with great effort, only spotty information can be provided which by its very nature lacks one quality of observation most necessary to synoptic meteorology, that of complete continuity in time and space.

A third possibility, however, has been suggested[1-4] that might indeed provide the necessary information about remote regions of the Earth. This is weather reconnaissance by means of satellites. The questions that one might ask about such a system are:

1. What are its limitations?
2. Within these limitations, what weather information is obtainable, and is it adequate to do the job?
3. What is the growth potential of a satellite weather reconnaissance system?
4. Can this information be fitted into the present weather analysis and prediction system?

Satellites as weather observing tools seem to offer a new dimension in the field of meteorology: area. That is, they would provide us with the ability to view the Earth as a whole in a very short period of time. As with any new tool, the ultimate value of a weather satellite must grow out of experience gained by using it. Initially, weather observation from a satellite will be obtained mainly by optical means, by simply looking down at the visible manifestations of weather. The first attempt to do this is part of the IGY satellite program, and William Stroud and his associates at the Signal Corps Engineering Laboratories have developed a scanning system for the Vanguard satellite.[5]

In this type of operation almost all of the regular quantitative measurements usually associated with synoptic meteorology tend to fall by the wayside. It will be impossible to do more than make an intelligent guess at the values of temperature, pressure, humidity, and other conventional meteorological parameters. However, some added information can be gathered if an ability to look at the Earth and the atmosphere in the infrared part of the spectrum is also provided.

The basic limitation of early weather satellites, then, is the degree to which meteorologists can utilize data that will be largely qualitative. Since clouds are the objects most readily discernible from extreme altitudes, they become the predominant item of observation for use in forming a weather picture.

While cloud data alone cannot tell everything about the current weather situation, it does appear that with theoretical knowledge and meteorological experience with satellite data, an accurate cloud analysis can produce surprisingly good results in areas where no other information is available. Further, in areas where good data are currently obtained from the surface, satellite cloud observations can provide continuity and completeness that are not given by the present weather-observing network.

A. Weather Information Obtainable from Early Satellites

THE VISIBILITY OF CLOUDS FROM SATELLITES

On the basis of a rather extensive body of evidence, it appears that except for the limited cases of snow background or water illuminated by the Sun low in the sky, clouds will have a brightness at least twice as great as that of the general Earth background.[6-9] A contrast of this extent between Earth and clouds means that clouds will be observable with television or photographic equipment in a satellite.

Cloud photography of useful quality can be obtained with an observation system exhibiting even rather poor resolution. For example, gross cloud cover can be obtained with a system than can resolve ground dimensions of the order of 1 mile or more from an altitude of several hundred miles. Study of cloud pictures taken from rockets, such as figure 1, indicates that it is necessary to resolve ground dimensions of the order of 500 to 1,000 feet in order to identify individual cloud types. Such resolution is attainable with current television and optical systems, although communication links of rather large capacity would be required to relay data to the ground if large

FIG. 1. Cloud picture taken from rocket

areas are to be viewed. Therefore, recognition and identification of clouds is readily feasible by satellite weather reconnaissance, but communication of the information rapidly to ground stations will be a limitation in first vehicles.

THE USE OF CLOUD OBSERVATIONS

Cloud observations from a satellite could provide the meteorologist with a view of the entire world weather pattern that he can hardly achieve by present indirect methods. In addition, some detailed information can be extracted from cloud photographs.[10] Wind direction may be estimated in several ways: (1) Present meteorological techniques have established that certain definite weather situations will produce certain sequences of clouds preceding or following them. Using mathematical models for preliminary orientation, the wind direction may then be approximated through a knowledge of the theoretical circulation associated with a given weather situation. (2) Cumulonimbus clouds may extend from as low as 1,600 feet up to 40,000 feet, and their slope becomes an indication of wind variation with altitude. (3) Use can be made of the fact that cumulus clouds form on the lee side of mountains. (4) The direction of movement of atmospheric pollutants such as industrial gases will indicate the direction of winds at low altitudes.

Temperatures may be estimated by starting with the statistical normal for the time of year. These preliminary estimates may then be modified by the various affecting conditions. Cloud systems, wind direction, and even forms of general ground cover (snow, etc.) will aid the analyst in deciding whether the area under observation is being affected by relatively cold or warm air. Upper air temperatures may be estimated in the same manner, clouds indicating the boundary between air masses (fronts). The slopes of vertically developed cloud forms will also aid in determining the temperature gradient of the surrounding area.

No quantitative measures of barometric pressure can be obtained from observations of the types under consideration. Furthermore, it now appears to be virtually impossible to make even a qualitative estimate, other than to determine whether the area is under the influence of a high- or low-pressure system.

INFRARED OBSERVATIONS

Infrared and other electromagnetic measurements provide powerful tools for investigating and observing our atmosphere. By proper choice of the regions of the spectrum to use for observation, it is possible to obtain temperatures at various altitudes in the atmosphere. The altitudes would actually be fixed as the tops of various layers of atmospheric gas that absorb radiation in different parts of the spectrum. Examples are the ozone layer at approximately 100,000 feet, and the region at about 40,000 feet where atmospheric water vapor becomes very tenuous. By very careful observation it may be possible to determine not only the temperatures at the tops of these layers, but also the quantities of various gases that exist in the atmosphere.

Measurement of atmospheric gas content would be of considerable value. The atmosphere can be likened to a large heat engine, with the Sun providing the power and the air acting as the working fluid. The reactions of the atmosphere, and the resulting visible manifestation that we call weather, depend on the amount of energy absorbed by the various gases that make up the atmosphere. Thus, the variations in the amounts of the various gases available are necessary parameters in any research leading to an improved ability to forecast weather. More immediately, the local moisture content or water-vapor content of the atmosphere is needed in forecasting the advent of clouds and precipitation.

It is also of interest to note that a comparison of the amounts of solar radiation entering and leaving the atmosphere will yield a better measure of the "heat balance" of the atmosphere. Better determination of the heat balance should shed new light on the way in which the Sun-atmosphere heat engine drives the massive circulation cells that are known to exist in the atmosphere and which serve to distribute energy around the Earth.

IMPROVED FORECASTING THROUGH SATELLITE OBSERVATIONS

It is not clear that the first meteorological satellites will markedly improve forecasting ability. However, one might expect at least a modest immediate improvement in the following ways:

1. Improved coverage and continuity over ocean areas, leading to improved ability to forecast for the western side of the North American Continent, and in air and shipping lanes in both oceans.
2. Improved coverage over areas where hurricanes and typhoons are born and grow, leading to possible improvements in hurricane and typhoon warning service.

Over a long period of time the greater continuity and total coverage should itself provide some measure of overall improvement in forecasting ability.

COVERAGE AND CONTINUITY

Since weather is a dynamic phenomenon, an improvement in forecasting depends to a large degree upon the opportunity to observe more or less continuously over a very wide area. For most weather phenomena, the maximum cycling time is of the order of 24 hours. Wide coverage within such a period of time is possible with observation satellites.

DISSEMINATION OF SATELLITE WEATHER DATA

To be useful, weather information must be quickly available to the various meteorological services. Its value decreases very rapidly with time after it is first obtained.

The maximum time from the instant of acquisition to actual dissemination should be held to within something like an hour. This rate of information flow suggests a rather elaborate pickup and relay system. One might visualize a "Satellite Weather Data Center" where incoming data would be processed to extract the maximum amount of usable meteorological information for immediate dissemination.

The problems of data analysis and dissemination warrant at least as much thought and effort as the design and operation of the data-gathering vehicle.

B. The Present Meteorological Satellite Program

The Advanced Research Projects Agency (ARPA) has initiated a meteorological satellite development program with the following main divisions: rocket vehicle, to place the satellite in orbit at an altitude of about 300 miles; satellite packaging, containing the sensors, storage, power, and radio communication equipments; ground tracking and data readout network; data handling and processing system; data analysis procedures. Parts of the complete system are classified for the present; however, a description of the data to be obtained has been released.[11]

The primary observations will be of the patterns of cloud cover, obtained by miniature vidicon television cameras designed by the Radio Corporation of America. One camera will sweep out a path some 1,200 miles wide and about 6,000 miles long. Ten pictures, 1,200 miles on a side, will be taken on each orbit revolution, and these will overlap to make the strip. Since there will be 500 television lines in each picture, the ground resolution will be about 2.5 miles. The cameras will be most sensitive in the red part of the spectrum in order to reduce the blue light which is scattered from the atmosphere. In addition, in order to obtain some cloud pictures with higher resolution for comparison, two other cameras with longer focal lengths will be provided. These will cover a smaller area on the ground, though the number of higher resolution pictures will be greater—34 pictures per orbit revolution with each camera.

The pictures are to be taken at a point in each orbit which is most suitable with respect to sunlight and satellite orientation. Pictures are then to be stored on magnetic tape and subsequently read out while the satellite passes over a ground station—a period of at least 4 minutes is required for the readout.

The transmitted pictures will be handled in two ways. On the ground the video signal will be stored on magnetic tape, and a direct television display will be photographed as the pictures are received. Thus, pictures can be available within minutes from the time of taking. The principal use of these pictures will be made after they have been rectified and located geographically by a central data-processing center.

A variety of infrared sensors will also scan the Earth and give a new and unique set of observations. These sensors have been designed by the Army Signal Research and Development Laboratories (ASRDL), and are a direct outgrowth of the IGY "weather satellite" built by ASRDL. From these observations, the heat budget of each area observed can be estimated.

The value of these infrared measurements, though still untested, is expected to be substantial.

The two agencies responsible for analysis and use of the data are the Geophysics Research Directorate of the Air Force's Cambridge Research Center and the United States Weather Bureau.

Preliminary experiments in meteorological observation from a satellite are planned in the Vanguard program with photocells as sensors in a total payload weight of about 20 pounds.[12]

C. Lines of Future Development

Future meteorological satellite developments, as indicated above, should include capabilities to measure cloud motion, atmospheric temperatures, total moisture content, total ozone content, and total radiation flow into and out of the atmosphere.

It must be emphasized that many of the ideas discussed here have not been completely proved by experiment. Enough information is at hand, however, to indicate clearly the desirability of employing satellites for weather reconnaissance.

The first weather satellites will be exploratory, and as more experience is gained in the interpretation of these radically new kinds of observations their usefulness should grow. It is conceivable that in the not so distant future, satellites may actually supplant a part of our present weather network.

In addition to direct experiments with meteorological satellites, certain IGY experiments should cast light on this area. Specifically, useful results should flow from the cloud-cover experiment of the Army Signal Engineering Laboratory and the radiation-balance experiment of the University of Wisconsin.

D. Weather Control

Much has been written and said in the past few years on the subject of weather control. The reason for such speculation, as much of it has been, is obvious when one considers the effect that weather and its vagaries have on our day-to-day living (both economically and physically). The ability to control the weather will not be easily acquired. Rather it will probably grow out of a better understanding of the basic processes in the atmosphere. From this understanding will come an indication of the chain of physical events that lead to a particular weather phenomenon. Not until we completely know and understand this chain and the part that each link plays will we have any hope of influencing weather. The part that space experiments can play at present in weather control is that of providing a research tool, such as the meteorological satellite, which will aid in gaining the necessary better understanding of the atmosphere.

Notes

[1] Greenfield, S. M., Synoptic Weather Observations from Extreme Altitudes, The RAND Corp., Paper P-761, February 15, 1956.

[2] Widger, W. K., and C. N. Touart, Utilization of Satellite Observations in Weather Analysis, Bulletin of the American Meteorological Society, vol. 38, No. 9, 1957, pp. 531-533.

[3] Wexler, H., Observing the Weather from a Satellite Vehicle, Journal of the British Interplanetary Society, vol. 13, 1954, pp. 269-276.

[4] National Aeronautics and Space Act, hearings before the Special Committee on Space and Astronautics, U. S. Senate, 85th Cong., 2d sess., pt. 2; H. Wexler, p. 369.

[5] Annals of the International Geophysical Year, vol. VI, pts. I-V, 1958, Pergamon Press, pp. 340-345.

[6] See footnote 1.

[7] Glaser, A. H., and J. H. Conover, Meteorological Utilization of Images of the Earth's Surface Transmitted from a Satellite Vehicle, Harvard University Blue Hill Meteorological Observatory, October 1957.

[8] Hewson, E. W., Quarterly Journal of the Royal Meteorological Society, vol. 69, 1943, p. 47.

[9] Hewson, E. W., and R. W. Longley, Meteorology, Theoretical and Applied, John Wiley & Sons, Inc., New York, 1944, pp. 73-75.

[10] See footnote 1.

[11] Minutes of the first meeting of the Committee on Meteorological Aspects of Satellites, Space Science Board, National Academy of Sciences, September 26, 1958.

[12] Satellites Will Advance Knowledge of Weather, Department of Defense, Office of Public Information, news release 871-58, September 26, 1958.

23

Navigation Satellites

Artificial satellites can provide the basis for all-weather navigation systems to determine with accuracy the position, speed, and direction of a surface vehicle or aircraft. Other methods, such as dead reckoning, Sun and star sighting during clear weather, and use of inertial guidance devices, can provide such information with adequate precision in many applications for limited periods of time. It is, however, necessary in many cases to check, independently and periodically, the navigation data indicated by these systems.

In checking and correcting navigational data, two distinct techniques employing a satellite are of interest: (1) sphereographical navigation, which is very similar to celestial navigation, and (2) a method which makes use of the doppler-shift phenomenon.[1-4]

A. Sphereographical Navigation

Celestial navigation employs measurement of the angle between the vertical and the line of sight to a celestial body. The position of the observer on the Earth's surface can be determined from a pair of such observations of two celestial bodies. The same kind of procedure can be used to determine position from two successive observations of a satellite. There are, of course, differences in detail between navigation by stars and by satellite, since star positions change very slowly while a satellite fairly close to the Earth moves at great speed.

In both applications, the observer must also determine the local direction of the vertical by pendulum or some other device.

For all-weather navigation, the satellite would radiate a continuous radio signal. The observer, equipped with an electronic sextant and an indicator of the vertical, would then be able to determine his position from radio observations of the satellite.

B. Doppler-Shift Navigation

An all-weather navigation system can also be based on a satellite broadcasting a continuous radio signal, by using the "doppler-shift" principle. The basic phenomenon is the following: The radio signal received from a moving vehicle will appear higher in frequency as the vehicle approaches the observer and lower as the vehicle recedes from the observer. The difference between the observed signal frequency and the known transmitter frequency is a measure of relative position and motion of the vehicle and observer; and, therefore, proper use of such frequency shift information can provide navigational data.

This type of system does not require determination of the local vertical by the observer.

C. Navigation Tables

For either of the above methods, the observer must know the true position of the satellite at the time of observation. Thus, he must be provided with a table of satellite positions covering the duration of his trip. These tables must be prepared in advance as mathematical predictions, as is now done by the Naval Observatory in the case of tables of positions of celestial bodies for navigation purposes.

Navigation accuracy is dependent upon the precision with which satellite position can be predicted into the future. This precision is, in turn, dependent largely upon the accuracy of observations, the computational procedure used, the accuracy with which the relevant physical constants are known, and the magnitudes of unpredictable disturbing effects acting on the satellite. The most important of these disturbances is the uncertain air drag at lower altitudes. Since it is at fairly high altitudes, the orbit of Vanguard I can be predicted into the future for about a month with reasonable accuracy, whereas predictions of the lower Explorer IV orbit are useful for only about 1 day into the future. Current predictions of Vanguard I position tend to be in error by some tens of miles after a month, but are in error by less than 5 miles over a few days. Prediction a few hours in advance would be off by less than a mile.

In addition to errors in satellite observations, there are two sources of difficulty in accurate orbit predictions. First, the classical methods used for many years by astronomers to determine orbits have not proven to be adequate when used in connection with artificial satellite orbits. New techniques or modifications of existing techniques appear to be necessary in order to improve the accuracy of orbital predictions. Second, further study must be made of the disturbing forces influencing orbital motion.

D. Equipment Requirements

The basic measurements required in sphereographical navigation are the azimuth and elevation angles of the satellite and the time of observation. The equipment of the navigator consists of a highly directional antenna and receiver, a clock, and equipment for defining the vertical. The accuracy of navigation is determined by various equipment characteristics. The factor with greatest implications for vehicle design is the antenna size.

As an indication of this size, an antenna about 12 feet in diameter will allow position determination within an error of about a mile when directed toward a satellite 1,000 miles away under representative conditions. A smaller antenna would lead to larger errors. The other equipments involved are generally comparatively small and light. The satellite would carry a transmitter.

For using the doppler technique of navigation the physical extent of the equipment needed on the navigated vehicle is less. The navigation equipment consists of a sensitive radio receiver, an accurate frequency reference, and an accurate clock. The equipment carried by the satellite would again be a transmitter, but one specifically designed to emit a very stable frequency.

Notes

[1] Lawrence, L., Jr., Navigation by Satellites, Missiles and Rockets, vol. 1, No. 1, October 1956, p. 48.
[2] Siry, I. W., The Vanguard IGY Earth Satellite Program, Naval Research Laboratory, presented to the Fifth General Assembly of CSAGI, held in Moscow July 30 to August 10, 1958.
[3] Leighton, R. B., Tracking an Artificial Satellite Using the Doppler Effect, California Institute of Technology, October 28, 1957.
[4] Quier, W. H., and G. C. Weiffenbach, Theoretical Analysis of Doppler Radio Signals From Earth Satellites, Johns Hopkins Applied Physics Laboratory, Bumblebee Series Rept. No. 276, April 1958.

24

Satellites as Communication Relays

Satellites can be used as components of a communication system to relay signals from one point on or near the Earth's surface to another.[1, 2]

Of singular interest for this purpose is the so-called "24-hour satellite" which completes 1 orbital revolution in 24 hours; so that if its motion is in the same direction as the Earth's rotation, and if other features of its orbit are properly selected, it will remain within line of sight of a fixed region on the Earth's surface.

Actually, the subsatellite point could remain truly fixed only if this point were on the Equator, and if the satellite orbit were perfectly circular, perfectly in the equatorial plane, and with precisely the right altitude (approximately 22,000 miles above the Earth's surface.) Even then, natural perturbations would cause a slight motion of the subsatellite point. Although these conditions cannot be fulfilled with perfect precision, reasonable precision in establishing the orbit will result in the subsatellite point's moving relatively slowly over a pre-selected area on the Earth's surface, and occasional adjustments with small on-board propulsion units can compensate for such defects indefinitely.

The advantages of a 24-hour orbit for a satellite as a communication relay are twofold:

> One such satellite would be within line of sight of nearly half the Earth's surface at any given time. Three properly placed satellites could provide virtually complete global communication coverage at all times.[3]
>
> The satellite would have a slow apparent motion relative to a point on the Earth's surface, thus simplifying the problem of properly pointing ground station antennas.

Some disadvantages relative to lower orbits are:

> It is more difficult to establish any given payload in a 22,000-mile-high orbit than to establish the same payload in a lower orbit.
>
> The power required to transmit signals between Earth and satellite is greater than for a lower satellite.

Satellites having lower orbits, and hence not remaining over a fixed area on the Earth's surface, can also be used as communication relays. The apparent motion of such satellites with respect to the ground would be more rapid than that of a 24-hour satellite, but could still be sufficiently slow to make tracking completely feasible. (For example, a 4,000-mile satellite would pass from horizon to horizon in roughly 95 minutes.)

The chief disadvantage of lower satellites is that they would be, at any given time, within line of sight of a smaller area on the Earth's surface. For example, a 4,000-mile satellite would be within line of sight of only about 25 percent of the Earth's surface at any one time. Thus, at least four such satellites would be needed to provide worldwide coverage. Actually, taking into account the details of the orbital motion, it turns out that the number necessary to insure worldwide coverage at all times would be greater than four, the exact number depending on the precision with which the orbits can be established. Low-altitude satellites may use data storage devices, such as tape recorders, to retain data received over one point for retransmission later over the intended point of receipt.

Communication satellites may be active or passive. An active satellite has transmitting equipment aboard, such as a transponder—a device which receives a signal from Earth amplifies it, and retransmits the same signal back to Earth (either immediately or after a delay). A passive satellite merely reflects or scatters incident radiation from the Earth, a portion of the radiation being reflected or scattered back in the direction of the Earth.

Passive satellite relays would require surface transmitters of much greater power than would active relays (unless the passive reflectors are extremely large); however, active satellite relays must carry aboard receiving and transmitting equipment and the necessary power sources, thus decreasing reliability and longevity.

An active satellite can be provided with omnidirectional transmitting antennas (radiating roughly equally in all directions) or directive antennas (radiating most of the energy toward the Earth). Directive antennas would require much less transmitted power, thus saving weight in that part of the payload devoted to transmission equipment and power supply, but would also require antenna stabilization so as to direct the radiated energy toward the Earth, thus increasing the payload weight devoted to attitude stabilization and its power supply.

A passive satellite relay could consist of an omnidirectional scatterer such as a spherical body, like a balloon satellite, or a directive scatterer such as a corner reflector. A corner reflector has the advantage that it tends to reflect radiation in the approximate direction from which the radiation came.

To give numerical examples can be misleading, since numerical results depend sensitively on the specific values assigned to such quantities as transmitted power, antenna sizes and directivities, operating frequencies, orbital characteristics, and other things. However, if reasonable values of these factors are assumed, an active relay could reasonably be expected to require about 0.10 watt of transmitted power (from the satellite) per kilocycle of channel bandwidth, assuming an omnidirectional satellite antenna. (A voice channel requires 5 to 10 kilocycles of bandwidth.) On the other hand, for a passive relay to have approximately the same communication capacity as an active relay which radiates, say, 10 watts, a spherical scatterer of the order of a mile in diameter would be required, or a corner reflector perhaps a few hundred feet in diameter. Such sizes do not preclude the use of passive relays, however, since it might be possible to construct such objects, or their equivalent in scattering power, with surprisingly little weight. The above numbers could easily change by factors of ten or even one hundred if different values of relevant parameters were assumed.

Perhaps the most important question to consider is that of the value of satellites as components of communication systems, both civil and military. This question is generally one of economics. The eventual feasibility of establishing satellites as communication relays cannot be doubted. On the other hand, alternative methods are available for relaying signals long distances, say around the world, over either land or water. Among these alternative methods are high-frequency radio; very-high or ultrahigh frequency wireless communication using land-based, shipborne, or airborne relay stations; transmission lines and cables (including submarine cables); and tropospheric, ionospheric, or meteor-burst scatter techniques.[4] Establishing the usefulness of satellite relays would involve a thorough analysis of the comparative cost of using satellites as opposed to (but, of course, not necessarily to the exclusion of) other available alternatives, for specified levels of capacity and reliability.[5]

In assessing the military usefulness of satellites as communication relays, the same sort of cost comparison must be made, taking into account possibly different required levels of capacity and reliability as well as additional factors such as security and vulnerability.

Even in cases where satellites may not be economical in direct dollar cost, they may be very valuable because they can be set up quickly. Extensive long-distance communication systems in some forms take years to complete—a satellite can be launched in a very short time.

Satellites can provide radio broadcasting coverage of wide areas with the use of several interconnected ground transmitters.

Satellites can be used to pick up messages transmitted

straight up and rebroadcast them straight down at the intended point of receipt even if that point is moving, as, for example, a ship. Low power levels may be used throughout, and interference among neighboring stations on the same frequency can be eliminated. This is a point of considerable importance, since it would greatly increase the number of stations that could use a given region of the radio-frequency spectrum.

Notes

[1] Adams, C. C., Space Flight, McGraw-Hill Book Co., Inc., New York, 1958, p. 278.

[2] Haviland, R. P., The Communication Satellite, presented at the Eighth International Astronautical Congress, Barcelona, October 6-12, 1957.

[3] Clarke, A. C., Extraterrestrial Relays, Wireless World, October 1945.

[4] Chu, Ta-Shing, Ionospheric Scatter Propagation at Large Scatter Angles, Antenna Laboratory, Department of Electrical Engineering, Ohio State University Research Foundation, July 1, 1957.

[5] Pierce, J. R., Orbital Radio Relays, Jet Propulsion, vol. 25, No. 4, 1955, p. 153.

25

Balloon Satellites

There are a number of scientific reasons for wishing to have an easily visible satellite.[1] For example, precise orbit determination will probably be done optically, and it is clearly desirable to have a satellite which reflects or emits a considerable amount of light. An available method is the creation of a large reflecting object such as a metal balloon.[2]

Orbit determinations based on optical tracking can be used to:

Determine air drag at high altitudes, from which atmospheric density can be derived. A possible complication is the effect of an electrostatic charge on the satellite, and the interactions between this charge, the ions present in the ionosphere, and the Earth's magnetic field.

Make geodetic measurements of the size and shape of the Earth.

Measure ion densities around the Earth, when coupled with certain precision radio techniques.

Metal balloon satellites also have potentially important military applications. Such an object, of very light weight, can be made in a size suitable to simulate an ICBM nose cone. It could, therefore, provide a permanent, realistic target for development and training activities in support of ICBM defense systems.[3] Lightweight metal balloons simulating ICBM nose cones can also serve as decoys to frustrate ICBM defense systems.[4, 5] Correspondingly, one would expect balloons, as decoys for military satellites, to be effective in protecting satellites against attack; or balloon satellites permanently on orbit could be made numerous enough to "saturate," on a long-term basis, ICBM warning radars.[6]

Balloon satellites that can simulate weapons of military offense suggest some potentially serious implications for warning systems and retaliation policy. For example, such an object could take up a trajectory similar to that of an incoming ICBM as a result, say, of a faulty satellite launch attempt. Further, such a faulty launch attempt could, in some circumstances, lead to a situation having the appearance of a mass attack, since a large satellite payload capacity could be used to carry a great many metal balloons. To illustrate, the approximately 3,000-pound capacity of the Sputnik III launching system could be used to carry in 1 package as many as 300 of the 12-foot balloons constructed for Explorer IV, since each of these balloons weighs less than 10 pounds.

Notes

[1] Astronautics and Space Exploration, hearings before the Select Committee on Astronautics and Space Exploration, 85th Cong., 2d sess., on H. R. 11881, April 15 through May 12, 1958; Technical Panel on the Earth Satellite Program, National Academy of

Sciences, Basic Objectives of a Continuing Program of Scientific Research in Outer Space, p. 782.

[2] NASA Earth Satellite, fact sheet 1, National Aeronautics and Space Administration.

[3] Astronautics and Space Exploration, hearings before the Select Committee on Astronautics and Space Exploration, 85th Cong., 2d sess., on H. R. 11881, April 15 through May 12, 1958; Maj. Gen. J. P. Daley, p. 688.

[4] Nike-Zeus May Be Inadequate, Top Defense Scientist Warns, Aviation Week, vol. 69, No. 19, November 10, 1958, p. 33.

[5] Address by Dr. Richard D. Holbrook, Advanced Research Projects Agency, Institute for Defense Analysis, to the Atlanta professional chapter, Sigma Delta Chi, Atlanta, Ga., September 9, 1958.

[6] Inquiry into Satellite and Missile Programs, hearings before the Preparedness Investigation Subcommittee of the Committee on Armed Services, United States Senate, 85th Cong., 1st and 2d sess., pt. I; W. von Braun, p. 602.

26

Bombing from Satellites

The feasibility of delivering devices from a satellite to the Earth's surface implies the possibility of bombing from a satellite, note of which has been taken in both the United States and U.S.S.R.[1-6]

The extent of the propulsion installation required on the satellite to eject the warhead depends upon the distance to be covered by the warhead during its descent phase. If the warhead is to descend essentially straight down from directly above the target, the propulsion requirement is very severe, indeed. However, the problem quickly becomes an entirely reasonable one if distances of several thousands of miles are allowed for accomplishing descent, as illustrated in figure 1. If the descent range is, for example, 5,500 nautical miles, descent from a 300-mile orbit can be accomplished by reducing warhead velocity by less than 1,000 feet per second. This is the approximate speed of a projectile from a 75-millimeter field gun, with reduced charge, to cover about 7,000 yards.[7]

Some additional propulsion capacity would be needed to deflect the warhead to the right or left, since a prospective target will rarely lie directly under the satellite orbit path.

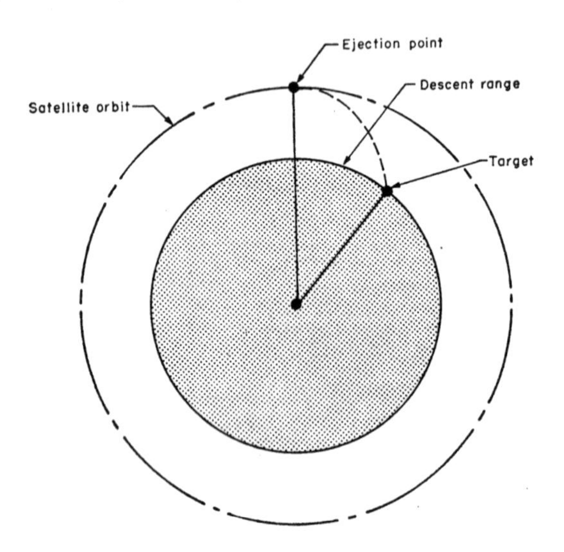

FIG. 1. Bomb delivery from a satellite

For unmanned satellites, the descent can be initiated and controlled directly from a ground radar station in sight of the ejection point; or, alternatively, ejection from orbit and descent guidance could be carried out on a fully automatic basis in the satellite at any time after receipt of a general signal from the ground to proceed to the attack. The accuracy of fully automatic delivery would depend upon the performance of internal equipment, as well as upon orbit data acquired from time to time from ground installations. A critical factor is the need for very precise knowledge of attitude direction references in space. For example, an error of one one-hundredth degree in azimuth reference will cause a strike error of 1 nautical mile for a 5,500-nautical-mile descent range.

For such a bombing system, satellite launchings could be conducted long in advance of a war, with any desired degree of leisure in a completely peaceful environment. Propellants used in the launching rockets could be chosen for maximum payload performance, and launchings could be conducted from one or a few sites located favorably with respect to weather, population density, convenience to production and supply sources, etc.

These features of convenience must be balanced against the fact that the rather extensive installation in a bombing satellite on orbit must operate reliably for some long period, if the replacement rate is to be held to a supportable level. Development of space-vehicle equipment with a very long reliable life is a basic necessity for a great many astronautics activities—working life spans of the order of a year or more would be representative of the requirements for interplanetary flights, various satellite missions, etc.

Bombs could also be delivered from manned satellites; for such a case, the guidance operation could include direct line-of-sight steering of the bomb-carrying missile to the target—even a moving target.

Notes

[1] Gavin, Lt. Gen. J. M., War and Peace in the Space Age, Harper & Bros., New York, 1958.

[2] Pokrovskii, Maj. Gen. G. I. (Red Army), The Role of Science and Technology in Modern War, The RAND Corp., Translation T-79, February 5, 1958; translated by H. S. Dinerstein.

[3] Inquiry into Satellite and Missile Programs, hearings before the Preparedness Investigating Subcommittee of the Committee on Armed Services, United States Senate, 85th Cong., 1st and 2d sess., pt. I; W. von Braun, p. 601.

[4] Leghorn, R. S., Warfare, Stalemate and Security, an address presented at the Franklin Institute, December 18, 1957.

[5] Statement by Senator Clinton P. Anderson in the United States Senate, May 6, 1958.

[6] Compilation of Materials on Space and Astronautics, No. 2, Special Committee on Space and Astronautics, United States Senate, 85th Cong., 2d sess.; C. Holifield, p. 100.

[7] Firing Tables for Gun, 75-mm, M 1897, Firing Shell, H. E., M 48, Ordnance Department, U. S. Army, FT 75-Z-2, 1940.

27

Scientific Space Exploration

Most of the information that man is able to obtain about the universe comes to him in the form of electromagnetic radiations. Some of these radiations he can see with the naked eye. But visible light represents a very small part of the total radiation spectrum (see figure 1); the Earth is constantly being bombarded with other forms of "light," which are invisible to the human eye.

If we spread out the entire radiation spectrum as it occurs in nature, we find that a star like the Sun concentrates most of its energy in a relatively narrow band stretching from the near ultraviolet into the infrared portion of the spectrum. That is, most of the Sun's energy is visible to us by optical means. In addition, however, very high energy radiations of short wavelength in the form of X-rays, gamma rays, and cosmic rays are present, as well as radiations of low energy and long wavelength, all of which are invisible to the eye. The eye is sensitive only to those radiations that fall within a small band of wavelengths. Radiations whose wavelengths are longer or shorter than those to which the eye can respond must be "seen" with other sensors.[1]

We can see into these hidden parts of the spectrum by using various kinds of detectors. Properly coated photographic plates, for example, allow us to look part of the way into the infrared region, and they also permit us to look into the ultraviolet, X-ray, and gamma-ray regions of the spectrum. Other devices, such as photoemissive cells, photoconductors, and bolometers, allow us to peer farther into the infrared, while Geiger counters, scintillation counters, etc., permit us indirectly to "see" gamma radiation and some of the high-energy cosmic rays that come through the atmosphere. Radio antennas, of course, provide us with means to look into the longer wavelength portion of the radiation spectrum.

While we have developed sensors that will enable us to see and measure much of this very short and very long wavelength energy, we have not been able to exploit them fully because only a part of the energy in these portions of the spectrum reaches the Earth. Much of it is blocked or deflected, absorbed or screened out by the Earth's atmosphere and magnetic field.

Atmospheric turbulence, air currents, and eddies make the images of the celestial bodies shimmer and dance about on the photographic plates of telescopes, so that we can only see a blur where we should see a sharp picture. The biggest telescopes on Earth normally operate with an effective resolution which is about one-twentieth of the theoretically obtainable value.

Low-energy radio waves are reflected and absorbed by the electrons and ions of the ionosphere, and never get through the atmosphere. In fact, most of the radio-frequency energy that envelops the Earth cannot penetrate the ionosphere.

The Earth's atmosphere is completely opaque also to wavelengths shorter than a few millimeters, and it does not "open up" until one reaches the near infrared, where ordinary light and heat rays exist. At still shorter wavelengths, because of the influence of ozone and other atmospheric gases, the Earth's atmosphere blots out the ultraviolet radiation from the Sun and the other Stars. The so-called soft X-rays emitted by the Sun can only penetrate the outermost layers of the Earth's atmosphere. The atmosphere in general is opaque to X-rays and to gamma rays. Only the most energetic cosmic rays succeed in penetrating the Earth's atmosphere down to the ground, so that information about the lower energy primary cosmic rays is not available on the surface of the Earth. Moreover, the Earth's magnetic field repels low-energy primary cosmic rays and is another factor in keeping them from reaching the ground.

In fact, the plight of the Earth-bound observer until now may be likened to that of a man imprisoned in a stone igloo with walls and roof 10 feet thick. At the time of his imprisonment, a small hole in the roof directly overhead was his only connection with the outside world. Since then, through a great deal of labor, aided by ingenuity, he has been able to enlarge

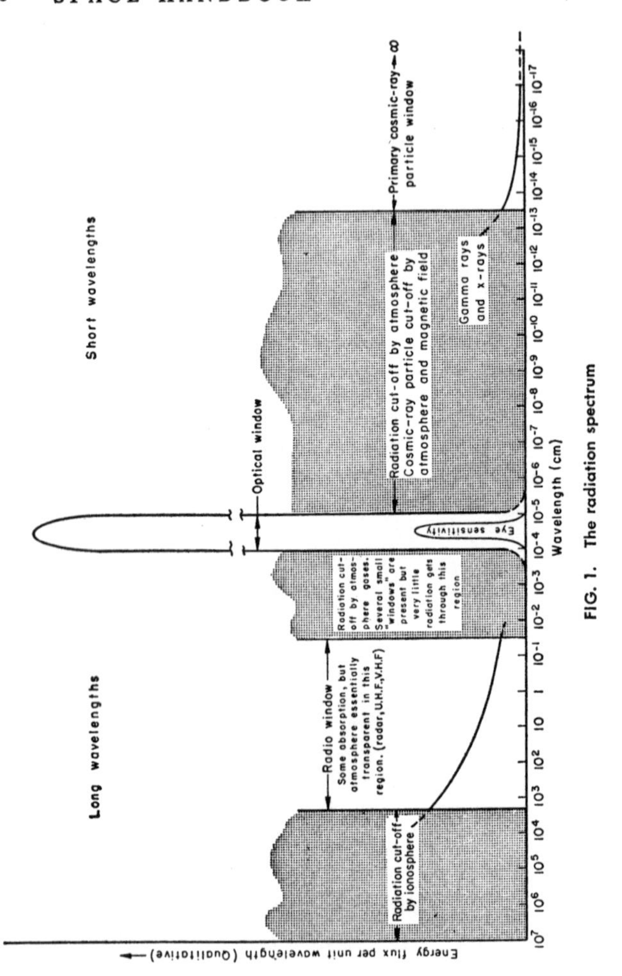

FIG. 1. The radiation spectrum

this roof opening slightly, and also to bore two small holes in the wall close to the ground, on opposite sides of his prison. Through these three "windows" comes all the information that he is able to obtain about the world around him, and it is with this meager knowledge, coupled with his imagination, that he must construct a picture of the outside world.

Applying this idea to the radiation spectrum in figure 1, we may say that man's knowledge of the universe has come to him through three "windows" in the spectrum. Through a radio window he "sees" a wide range of radio frequencies from different types of celestial objects: the Sun, the planets, possibly other stars, clouds of gas in our own galaxy, exploding stars, and outside universes. Through an optical window, he has been able to obtain other information about the universe with the aid of telescopes, spectroscopes, cameras, etc. And finally, there is the cosmic-ray window, through which still other bits of information have come to him. Except for these apertures, however, the wall of the atmosphere shuts him in.[2]

A. Fundamental Problems

A prime question of concern to astronomers is the composition of the universe; that is, the relative abundance of different species of atoms. The most abundant elements in the universe are thought to be the lighter ones: hydrogen, helium, carbon, nitrogen, oxygen, neon, etc. Unfortunately, their strongest spectral lines are in the far ultraviolet region of the radiation spectrum, which can never be observed from the surface of the earth. While indirect methods have given us some preliminary data, the fundamental questions of how much of these substances exists and whether the stars differ in composition from one another must still be considered as unsolved. The answers would be of the utmost importance to astronomy. These spectra, especially in the bright and very hot stars, could be observed with apparatus in the few-hundred-pound satellite-payload range, and the results telemetered to earth with equipment that is now available. The total amount of information that must be transmitted to obtain the first spectroreconnaissance of the stars is in fact quite small, and useful observations could be done within today's technology.

Obviously, the nearest star, the Sun, is worthy of the most detailed study. This is true not only because the Sun is the prime source of the Earth's energy, but because the ultraviolet, X-ray, and gamma-ray radiation of the Sun will be controlling factors of man's environment in space, and they may provide hazards to man's survival which are of a nature still unknown and unsuspected.

Moreover, the effect of the Sun's ultraviolet radiation dominates the ionization of the outermost atmosphere of the Earth, which has an important effect on communication, and possibly a very much more important one on the weather. The gain for pure scientific research, however, that could be obtained through detailed observation of the solar ultraviolet spectrum will come mainly in explanations of the origin of the hot outermost layers of the Sun. While the surface of the Sun is at a temperature of about 6,000° C., the temperature rises as one goes out from the Sun, reaching at least a million degrees at a distance of some 20,000 kilometers from the Sun's surface. We have no explanation of the origin of this heating; we do not know where the Sun's outer envelope stops, and there is a good chance that the Earth itself is immersed in this corona.

The Sun is subject to violent storms, manifested by so-called sunspots, prominences, and solar flares. These disturbed outer regions of the Sun have an important effect on the Earth's ionosphere, geomagnetism, and, in the long run, ordinary weather. It is technically feasible to map the Sun in terms of its emissions in the ultraviolet or even soft X-ray portions of the spectrum. Such observations, unobstructed by the atmosphere, would provide us with a detailed history of what happens during solar storms and flares. Pictures could be telemetered back on a more-or-less continuous basis, providing warning of solar disturbances which could effect the Earth.

To obtain a true picture of the total cosmic-ray energy enveloping the Earth, particularly that which reaches us from the Sun during solar storms or flares, it will be necessary to reach beyond the atmosphere and through the Earth's magnetic field to a distance of perhaps 25,000 miles. The longer wavelength primary cosmic radiation, where in fact most of the cosmic-ray energy is concentrated, must be investigated experimentally if we are to obtain information about the origin of these particles, their effect on the Earth and its inhabitants, and the hazards they may present to space travel.

The question of photographing other objects in our universe from space, and the possibility of our finding new kinds of objects, is an important one. Even low-altitude orbiting satellites, and certainly high-altitude satellites, would permit us to make detailed observations of the structure of the surfaces of the planets and the composition of their atmosphere. At the present time there are indications from ultraviolet measurements made from Naval Research Laboratory rockets that there are in space large clouds of gas of an unknown nature which shine in the ultraviolet. A photograph of even minimum resolution taken in the far ultraviolet would decide this question. In particular, astronomers have found that the most abundant element in the universe, hydrogen, is present throughout our Milky Way. This is detectable so far by radio observations only, although it is known that perhaps 10 percent of the total mass of our universe is in the form of gaseous hydrogen in space. Direct photographs, or spectra, of this interstellar hydrogen can be obtained once a vehicle has penetrated the atmosphere of the Earth, photography being important both in the ultraviolet and in the infrared.

Another very important question concerns the distribution of other systems of stars as one goes farther outward into space. Because of the apparent expansion of the universe, there is a shifting of the light emitted by distant objects toward the red portion of the spectrum, known as the "red shift." At the present time, external systems of stars, extragalactic nebulae, have been photographed out to distances of at least 2 million light years, and as far as we know, exist very much farther out into space. However, at great distances the red shift becomes larger and all the light of the system is shifted into the nonphotographable far infrared. There is no doubt of this effect, and what we do not know is how far out this red shift can be extrapolated. In particular, it is not necessary to photograph individual galaxies of very great faintness, but an important set of conclusions on the nature of the expansion of the universe and fundamental cosmology could be reached by merely measuring the total brightness in the far infrared of all of the galaxies together. Apparatus providing rough spectral resolution, so that quantitative measures of the brightness of the sky at many different wavelengths in the far infrared could be obtained, would perhaps settle some of the fundamental questions of the expansion of the universe, and the distance to which it stretches. This has an important bearing also on the age of our universe.

The subject of radio astronomy, which has grown rapidly in importance in the last few years, has already provided many scientific surprises. For example, the second brightest source of radio waves from outside the Earth turns out to be a pair of colliding galaxies at a distance of 300 million light years. Since most of the radio-frequency energy that occurs in nature does not penetrate the ionosphere, it can be measured only by probing above the ionosphere. Such investigations would greatly extend our knowledge of the total energy involved in radio emissions from these strange sources, perhaps giving us a clue to their origin.

One very important possibility considered by several scientists in recent years is that the existence of both extremely intense radio-frequency radiation and very high energy cosmic rays is an indication that we have still to discover some fundamental properties of the universe. One may speculate that the most fundamental processes in the universe are those involving extremely high energy particles, and that these may be produced by some as yet unknown physical mechanism. While tentative explanations of the origin of cosmic rays exist and suggest that they are merely matter accelerated to ultrahigh velocities, differing negligibly from the velocity of light, it would seem desirable to obtain a complete survey of the total electromagnetic and particle spectra at altitudes where the Earth is no longer a disturbing factor. Theories of the origin of the Earth have suggested that all matter was produced in a primeval explosion some 10 billion years ago. Are there any relics of this explosion available still in the form of very high energy radiation? Other theories suggest that matter is continuously produced in intergalactic space. Is this latter theory tenable? Are there any evidences in the radiations coming from space of the continuing creation of matter?

B. Experimental Programs

PRELIMINARY CONSIDERATIONS

Before mounting a large-scale attack on the space frontier, it is essential that we consider the interactions among possible experiments to be sure they are done in the proper sequence. It is conceivable, for example, that an early experiment, done merely because the means were available, could so alter the natural environment that other important experiments would no longer be possible. Experiments must be planned with due regard for leaving future parts of an overall program intact.

It must also be borne in mind that expensive scientific ventures in space will only be effective if backed up by adequate theoretical studies and laboratory research on the ground.

EXPERIMENTS IN SPACE BIOLOGY

Astronautics will provide new approaches to some of the fundamental problems in biological science. The study of terrestrial life forms in radically new environments (and perhaps even nonterrestrial life forms) will become possible, providing opportunities for increased understanding of the nature of the life process, how it originated, how it evolves and functions, and what forms it may assume under widely different environmental conditions. An experimental program in space biology should include:

Experiments to investigate the survival of micro-organisms under various atmospheric and space conditions.

Experiments to determine the astrophysical properties[3] of micro-organisms.

Actual samples taken at various levels of the atmosphere and exosphere to determine the presence (if any) of micro-organisms; and, later, samples of the atmospheres and soils of other bodies of the solar system.

Survival studies should be conducted both in the laboratory and in space. The laboratory experiments would entail subjecting bacteria, viruses, spores, and seeds to space-simulated conditions to determine the survival limits of organisms irradiated by X-rays, gamma rays, ultraviolet rays, and high-energy particles; and to determine the influence of temperature, absence of atmosphere, moisture, etc., on survival.

These laboratory tests should, of course, be supplemented by investigations under actual conditions in free space.

EXPERIMENTS WITH SOUNDING ROCKETS

Vertical rockets, of the Viking and Aerobee type, will continue to be useful, as in the past, for measurements of upper atmosphere phenomena and composition. They can also be used for high-altitude observations of the Sun, Moon, planets, and other celestial objects. (Balloons, of course, are also useful in this connection.)

EXPERIMENTS WITH SATELLITES

Uses of satellites for scientific observations have been mentioned under "Observation Satellites" and "Meteorological Satellites."

Satellites can remain in space permanently for long-term observation to altitudes of about 1 million miles, and would be most useful for continued mapping of the "radiation belt" disclosed by the IGY satellites.

Some experiments appropriate to satellites of various payload classes are:

1. Less than 100 pounds:

   Cosmic-ray counters; mass spectrometer; measurements of total magnetic field, solar ultraviolet radiation, X-rays, and gamma rays.

   Observations of heat balance and albedo of Earth.

   Low-resolution pictures of Earth, Moon, and planets.

   Density, composition, and temperature in upper atmosphere and solar corona.

   Geodetic information from radio and visual observations of satellite orbits.

   Micrometeorite densities.

   Ionospheric density from radio signals.

   Growth of spores or yeast in space.

   Small mammals in a space environment.

2. One hundred pounds:

   Small recoverable payload.

   Higher resolution photographs of Earth, Moon, and planets; emulsion plates for detecting cosmic rays.

   Measurement of vector components of geomagnetic field and its time variations.

   Spectrum of the atmosphere of the Earth and Sun; hemispheric cloud observations.

   Location of radiating points in space; cosmic radio noise at the low radio frequencies.

   Tests of space environment on mammals.

   Plant-growth cycle under weightlessness.

3. One thousand pounds:

   Telescope to obtain spectra of planets, stars, and galaxies from far ultraviolet through infrared.

   Combination meteorological satellite involving both television and infrared sensors.

   Test of the general theory of relativity using atomic or molecular clocks.

Small animals in space in a closed ecological system.

Experiments to note animal body rhythms and activity in space.

4. Ten thousand pounds:

   Larger animals and man in space.

   Large telescopes for astronomical observation.

LUNAR VEHICLES

Some of the purposes to which lunar rocket experiments might be turned include measurements of—

The mass of the Moon. The current estimates of this quantity may be in error by as much as 0.3 percent, and substantial inconsistencies exist between mass estimates based on asteroid observations and those implied by data on the motions of the Earth's polar axis.

The Moon's magnetic field: At present we have virtually no knowledge whatsoever of the Moon's magnetic field. Data on the magnetic field of the Moon would allow us to make some progress in theories about the history of the Moon, the processes of its formation, etc.

The composition and physical properties of the lunar atmosphere.

The composition and properties of the lunar crust.

Lunar surface temperature and its variation with time and depth.

Surface radioactivity and atmospheric electricity.

Seismic properties of the lunar interior.

INTERPLANETARY VEHICLES

The measurements that might be made on other planets are generally the same as those pertinent to the Earth itself, as modified by the singular features of each planet.

Notes

[1] We can, of course, produce many of these radiations, or forms of energy, on Earth. Radio and radar transmitters, light bulbs, X-ray equipment, and nuclear explosions all produce electromagnetic energy in different parts of the radiation spectrum. We cannot, however, as yet produce the highest energy cosmic rays, which are really solid particles moving at tremendous velocity, here on Earth.

[2] The areas beneath the curves in the three open regions indicate the amount of energy at various wavelengths that reaches the Earth. While the total quantity of energy that reaches us in the form of visible light is quite high, the amounts in the radiowave and cosmic-ray portions of the spectrum are relatively low. Most of these energies lie in a region of the spectrum to which we are denied access from the ground.

[3] By astrophysical properties are meant absorption coefficients, masses, sizes, etc.—in general, those properties which would give information on how the organisms would react to radiation fields and other forces affecting their transport and physical state in space.

PART 4

ASTRONAUTICS IN OTHER COUNTRIES

28

Astronautics in the U. S. S. R.

A. Soviet Interest in Space Flight

On December 21, 1957, the Central Committee of the Communist Party, Soviet Union, the Presidium of the Supreme Soviet of the U.S.S.R., and the U.S.S.R. Council of Ministers issued a proclamation noting the outstanding achievements that marked the 40th year of existence of the Soviet Government. The proclamation ended with the following announcement:

> For outstanding achievements in the field of science and technology, making possible the creation and launching of artificial Earth satellites, a large group of scientists, designers, and specialists have been awarded Lenin prizes.
>
> Scientific research organizations that participated in the development of the satellites and in the realization of their launchings have been awarded the Orders of Lenin and the Red Banner of Labor.
>
> For the creation of the satellites, the carrier rockets, the ground launching facilities, the measuring and scientific equipment, and the launching in the Soviet Union of the world's first artificial Earth satellites, a group of scientists, designers, and workers has been awarded the title of Hero of Socialist Labor. A large number of experts, engineer-technological workers, and workers have been awarded orders and medals of the Soviet Union. To mark the creation and launching in the Soviet Union of the world's first artificial Earth satellite it has been decided to erect an obelisk in Moscow, the capitol of the Soviet Union, in 1958.

Thus the official cloak of anonymity was draped collectively over the men and institutions that accomplished, with dramatic suddenness, man's first concrete step in the conquest of space. This announcement, in a sense, epitomizes the extreme caution with which the Soviets have handled the subject of rocket development since the mid-1930's when, after the first liquid-propellant meteorological research rockets were fired by Russian enthusiasts, the Soviet Government quickly realized the enormous military potential of the rocket and organized a Government-sponsored rocket research program, with its attendant security restrictions.

Russia has a rich historical background in astronautics that began at the end of the 19th century with the works of I. V. Meshcherskii on the dynamics of bodies of variable mass and the publications of K. E. Tsiolkovskii on the principles of rocket flight. Early Russian rocket enthusiasts made many fundamental contributions to this new technology.

Tsiolkovskii, the father of (and to the Soviets, the patron saint of) the science of astronautics, has been fairly well represented by rocket historians in western literature. Not so, however, his contemporaries—F. A. Tsander, who developed the idea of utilizing as fuel the metallic structural rocket ship components which were no longer necessary, and who in 1932 built and successfully tested a rocket motor operating on kerosene and liquid oxygen; Yu. V. Kondratyuk, who proposed the use of ozone as an oxidant and developed the idea of aerodynamically braking a rocket returning from a voyage in space; N. A. Rynin, who during the period 1928-1932 published a monumental 9-volume treatise on astronautics; Ya. I. Perel'man, the great popularizer of astronautics; and I. P. Fortikov, the organizer.

Until the door was shut on the publication of original material in 1935, rocket developments in the Soviet Union, especially those connected with the exploration of the stratosphere, were discussed quite freely. As early as 1929, an organization known as GIRD (after the initials of the Russian words for "group studying reactive motion") was formed by a number of scientists and engineers whose primary interest was in rocket engines and propellants. The papers written by various members of this organization contain a wealth of evidence of native competence in the various aspects of rocketry and space flight and clearly indicate that the Russians possessed a relatively high degree of technical sophistication more than two decades ago.[1] The GIRD publications included contributions by I. A. Merkulov, Yu. A. Pobedonostsev, and M. K. Tikhonravov, who are still very active in the field of rocket propulsion and space flight.

By 1929, V. P. Glushko, now a corresponding member of the U. S. S. R. Academy of Sciences, was already designing rocket engines, and from 1931 to 1932 he conducted test-stand firings with gasoline, benzene, and toluene as fuels, and with liquid oxygen, nitrogen tetroxide, and nitric acid as oxidants. The only liquid-propellant rocket engines mentioned by code designation in the Russian literature are the OR-2 engine, designed by Tsander, which in 1933 developed a thrust of 110 pounds operating on gasoline and liquid oxygen; the ORM (experimental rocket engine) series, designed by Glushko, of which the ORM-52 in 1933 developed a thrust of 660 pounds operating on kerosene and nitric acid; and the aircraft-thrust-augmentation rocket engines—RD-1, RD-2, and RD-3—which developed thrusts of 660, 1,320, and 1,980 pounds, respectively (1941-46). L. S. Dushkin designed an engine that propelled a meteorological rocket conceived by Tikhonravov to an altitude of 6 miles in 1935. Dushkin later designed an engine that developed a thrust of 330 pounds for a rocket plane (glider) built under the direction of S. P. Korolev and successfully flight-tested in 1940.

Appreciating the immense military potential of the rocket, the Soviet Government had organized, by 1934, a Government-sponsored rocket-research program—only 5 years after Germany had embarked on its rocket program, but 8 years before similar systematic Army-sponsored research began in the United States. Stalin's personal interest in the development of long-range, rocket-propelled guided missiles is discussed in the book *Stalin Means War* by Col. G. A. Tokaev, formerly chief of the aerodynamics laboratory of the Moscow Military Air Academy, who defected from the Soviet Zone of Germany to Great Britain in 1948.

After World War II, the Russians thoroughly and systematically exploited German rocket powerplants and guidance and control equipment; they then re-established the German state of the art as it had existed in 1945 by appropriating most of their rocket-test and production facilities and personnel. (It is interesting to note that, although most of the Germans were repatriated in 1952, a group of electronic experts was not repatriated until 1958.) The Soviets not only increased the thrust of the V-2 rocket engine from 55,000 to 77,000 pounds by increasing the propellant flow rate[2]—thereby extending the range of the missile from 200 to 700 miles—but also developed a super-rocket engine with a thrust of 265,000 pounds. They were also interested in designing a rocket engine with a thrust of 551,000 pounds, probably as an improvement on the powerplant the Germans had envisioned for their A-10 rocket. Events since August 1957 seem to indicate that the German A-9/10 project reached fruition in the Soviet ICBM.

These developments indicate that the Russian effort has been more than an extension of previous German work; to all indications it is based on independent thinking and research. This is not surprising, since Russia has its share of exceptionally capable technical men such as Semenov (the recent Nobel prize winner in chemistry) and Zel'dovich, Khristianovich, and Sedov—to mention but a few in the fields of combustion theory and fluid dynamics.

By 1949 the Soviets had embarked on an upper-atmosphere research-rocket program that involved the recovery by parachute, first of test-instrument containers and later of experimental animals. Papers dealing with this program were presented by S. M. Poloskov and B. A. Mirtov and by A. V. Pokrovskii in Paris in December 1956. According to a TASS report dated March 27, 1958, the single-stage rocket initially used (in May 1949) attained an altitude of 68 miles with an instrument payload of 264 to 286 pounds. With improved techniques, larger payloads were sent to higher altitudes. Thus, in May 1957, a single-stage geophysical rocket raised an instrument payload of 4,840 pounds to an altitude of 131 miles, while on February 21, 1958, an improved single-stage geophysical rocket raised an instrument payload of 3,340 pounds to a record altitude of 293 miles. In each case the payload was recovered by parachute. On August 27, 1958, a single-stage geophysical rocket launched in the Soviet Union reached an altitude of 279 miles with a payload of 3,720 pounds. Besides instruments for studying the upper atmosphere, the rocket carried two dogs in a special pressurized capsule. Both instruments and dogs were successfully recovered.

One of the meteorological rockets developed by the Soviets,

which has been used since 1950, was described in detail by A. M. Kasatkin at the Comité Spécial de l'Année Geophysique Internationale (CSAGI) Washington conference on rockets and satellites early in October 1957. It consists of a solid-propellant booster rocket 4.5 feet long and weighing 517 pounds that burns 180 pounds of powder in 2 seconds, and a 23-foot-long sustainer rocket having a starting weight of 1,495 pounds and a kerosene and nitric acid engine that develops a thrust of 3,000 pounds for 60 seconds. At an altitude of about 43 miles, the sustainer rocket separates into 2 parts, the upper part with instruments attaining an altitude of 50 to 55 miles. Both parts descend by parachute and are recovered.

The existence of an official Soviet space-flight program may be traced to a significant statement by Academician A. N. Nesmeyanov in his address to the World Peace Council in Vienna on November 27, 1953. Speaking on the problems of international cooperation among scientists, he said: "Science has reached a state when it is feasible to send a stratoplane to the Moon, to create an artificial satellite of the Earth." [3] As president of the U. S. S. R. Academy of Sciences, Nesmeyanov was, of course, familiar with all aspects of Soviet scientific progress; his statement clearly implied that Russian progress in rocket propulsion as of 1953 had made feasible such feats as launching an Earth satellite and flying to the Moon.

There is considerable evidence of early acceptance of the science of space flight by the Soviet hierarchy. It is not without significance that volume 27 of *Bol'shaya Sovetskaya Entsiklopediya* (*Large Soviet Encyclopedia*), published in June 1954, contained an article entitled "Interplanetary Communications" by M. K. Tikhonravov.[4] As of the end of 1957 there was no corresponding entry in any of the Western encyclopedias. Interestingly, *The New York Times* began to index articles on space ships and space flight under the term "Astronautics" only after the White House announced on July 29, 1955, that the United States intended to launch an Earth satellite.

Soviet interest in space flight was further revealed by the fact that on September 24, 1954, the Presidium of the U. S. S. R. Academy of Sciences established the K. E. Tsiolkovskii Gold Medal for outstanding work in the field of interplanetary communications, to be awarded every 3 years beginning with 1957. At about the same time, the Presidium established the permanent Interdepartmental Commission on Interplanetary Communications (ICIC) to "coordinate and direct all work concerned with solving the problem of mastering cosmic space." Academician L. I. Sedov, a topnotch hydrodynamicist, was appointed chairman, and M. K. Tikhonravov —who designed and successfully launched liquid-propellant atmosphere research rockets in 1934—was appointed vice chairman.

In addition to the ICIC, an Astronautics Section was organized early in 1954 in Moscow at the V. P. Chkalov Central Aeroclub of the U.S.S.R. Its goal was "to facilitate the realization of cosmic flights for peaceful purposes." Its charter members included Chairman N. A. Varvarov, Prof. V. V. Dobronravov, Design Engineer I. A. Merkulov, Stalin Prize Laureate A. D. Seryapin, Prof. K. P. Stanyukovich, Yu. S. Khlebtsevich, and International Astronautics Prize Winner, A. A. Shternfel'd.

Although the White House announcement of July 29, 1955 —that the United States intended to launch an Earth satellite sometime during the International Geophysical Year (IGY) —led to considerable speculation concerning the Soviet position and capability in this field of technology, the imperturbable Russians, as usual, did not commit themselves. Possibly they were only too well aware of the United States Earth satellite vehicle program, the existence of which was first publicly announced by Secretary of Defense Forrestal in December 1948.

A notable event occurred in the week following the White House announcement. The Sixth International Astronautical Congress sponsored by the International Astronautical Federation convened in Copenhagen, Denmark. It was notable because, unlike previous meetings, it was attended by two Soviet scientists, Academician L. I. Sedov, Chairman of the U. S. S. R. Academy of Sciences ICIC, and Prof. K. F. Ogorodnikov, a professor of astronomy at Leningrad State University, who was an exchange professor at Harvard in 1937.

The Russians were observers at the Congress and did not participate in any formal discussion of the papers. Sedov, however, did hold a press conference on August 2, 1955, at the Soviet Legation in Copenhagen, but unfortunately some of the statements attributed to him were garbled in the western press. Three days later, on August 5, *Pravda* published an official version of the press conference in which Sedov indicated that

> recently in the U. S. S. R. much consideration has been given to research problems connected with the realization of interplanetary communications, particularly the problems of creating an artificial Earth satellite. . . . In my opinion, it will be possible to launch an artificial Earth satellite within the next 2 years, and there is a technological possibility of creating artificial satellites of various sizes and weights. From a technical point of view, it is possible to create a satellite of larger dimensions than that reported in the newspapers which we had the opportunity of scanning today. The realization of the Soviet project can be expected in the comparatively near future. I won't take it upon myself to name the date more precisely.

Six months later, in February 1956, the Russians held a conference at Leningrad State University to discuss problems of the physics of the Moon and the planets. More than 50 scientists participated. The two principal topics for discussion were (1) the questions of planetology connected with the problems of astronautics and, primarily, the question of the state of the Moon's surface, and (2) the exchange of opinions and plans for observations of the coming great opposition of Mars in September 1956. Prof. N. P. Barabashev, conference chairman and director of the Khar'kov University Observatory, pointed out that the importance of planetology was growing substantially in connection with the demands of cosmonautics and that, at the same time, the responsibility of planetary, and especially lunar, investigators was increasing. M. K. Tikhonravov, Vice Chairman of the ICIC, enumerated the basic questions to which astronauts expect answers from the science of planetology.

At the Conference on Rockets and Satellites, held on September 11, 1956, during the fourth general meeting of the CSAGI in Barcelona, Spain, there occurred a prime example of official Soviet reticence to make factual pronouncements concerning rocketry and space flight. In presenting the general description of the Soviet Union's rocket and satellite program to an audience that was eagerly awaiting the Russian announcement, Academician I. P. Bardin, Chairman of the U. S. S. R. IGY National Committee and a vice president of the U. S. S. R. Academy of Sciences, read the following statement in Russian:[5]

> At the request of the General Secretary of the CSAGI, Dr. M. Nicolet, inquiring as to the possibility of the Soviet Union's participation in the rocket-satellite program, the Soviet National Committee announces that—
>
> (1) In addition to the U. S. S. R. program already presented to the Barcelona meeting the rocket-satellite program will be presented at a later time.
>
> (2) The U. S. S. R. intends to launch a satellite by means of which measurements of atmospheric pressure and temperature, as well as observations of cosmic rays, micrometeorites, the geomagnetic field and solar radiation will be conducted. The preparations for launching the satellite are presently being made.
>
> (3) Meteorological observations at high altitudes will be conducted by means of rockets.
>
> (4) Since the question of U. S. S. R. participation in the IGY rocket-satellite observations was decided quite recently the detailed program of these investigations is not yet elaborated.
>
> This program will be presented as soon as possible to the General Secretary of the CSAGI.

Needless to say, this unexpected and vapid statement left the assembled throng with a sense of complete frustration. No mention of it appeared in the Soviet press for more than 2 weeks. Finally, on September 26, 1956, a TASS report, captioned "Preparation for the International Geophysical Year" and bearing no dateline, appeared on page 4 of *Krasnaya Zvezda* (*Red Star*). The report quoted Academician Bardin as saying that

> The Soviet delegation's statement that work is being conducted in the U. S. S. R., just as in the United States of America, on preparations for upper atmosphere research by means of rockets and artificial satellites evoked great interest among the participants of the session. These satellites will revolve around the Earth, making a complete revolution in less than one hour and a half. They will be relatively small, approximately the size of a [soccer] football. They will weigh about 9 kilograms. Now scientists are making more precise a number of conditions for successfully launching the satellites. . . .

The Bardin item is in keeping with the generalized nature of Soviet reports and articles concerning their satellite plans and specifications. Stereotyped statements and reports are apparently a matter of policy. This was admirably summarized by an American IGY scientist, who (according to an Associated Press dispatch datelined Washington, October 2, 1957) said that Russian delegates had told him repeatedly that they consider it

> bad taste to make announcements in advance. Our policy is not to release any details until we have experimental results.

By 1956 the U. S. S. R. Academy of Sciences felt the need to apply for membership in the International Astronautical Federation. The application was voted on favorably during the Seventh International Astronautical Congress in Rome in September of that year. Moreover, the Soviet Union's lone observer-delegate to that Congress—L. I. Sedov—was elected a vice president of the Federation.

More than a year passed, however, before the Soviet Union complied with the bylaws of the International Astronautical Federation and submitted, through Sedov, a description (that is, an equivalent of a constitution) of the Academy's ICIC and a list of its members.

The main purpose of the commission, it seems, is to assist in every way possible the development of Soviet scientific-theoretical and practical work concerning the study of cosmic space and the achievement of space flight. Its specific duties and functions are manifold and involve the initiation, organization, coordination, and popularization of the problems of space flight, as well as the propagandization of the successes achieved.

The list of 27 members of the ICIC is a very impressive one. It includes eight academicians, some of Russia's—and the world's—top scientists. There is no question of the stature in world science of such men as P. L. Kapitsa, the famed physicist, N. N. Bogolyubov, the mathematical genius who is said to be the Russian counterpart of the late John von Neumann, V. A. Ambartsumyan, the noted Armenian astrophysicist, and others. Although most of the members of the Commission are pedagogues, that is, connected with some institute of higher learning, a number of them wear several hats, including military hats. Academician A. A. Blagonravov, for example, is a lieutenant general of artillery and is a specialist in automatic weapons. G. I. Pokrovskii is a major general of technical services and an explosives expert. V. F. Bolkhovitinov holds the rank of major general and is a professor of aeronautical engineering at the Military Air Academy. Yu. A. Pobedonostsev is a colonel, a professor of aerodynamics at Moscow State University, and a specialist in gas dynamics. It is quite evident that the military is well represented in the ICIC.

Because of the Soviets' extreme reluctance to reveal their activities in the field of astronautics, the myriad articles on the problems of space flight that appeared in the popular press prior to the end of 1956 presented, for the most part, well-known information from the western press with only occasional broad hints as to developments in the Soviet Union. Soviet technical journals, however, continued—as they had in the past—to present articles of considerable interest and merit, especially in the fields of flight mechanics and hydrodynamics.

Soviet uncommunicativeness ended in December 1956, when a delegation of 13 scientists, headed by Academician A. A. Blagonravov, an armaments specialist and a member of the Presidium of the Academy of Sciences, attended the First International Congress on Rockets and Guided Missiles in Paris. There the Russians presented two papers which revealed the prodigality of their rocket-test program: In the Soviet experimental technique, the measuring instruments are not carried in the rocket itself but in automatically jettisoned containers, the results being recorded on film and the containers recovered by parachute. The papers were entitled "Study of the Upper Atmosphere by Means of Rockets at the U. S. S. R. Academy of Sciences," by S. M. Poloskov and B. A. Mirtov, and "Study of the Vital Activity of Animals during Rocket Flights into the Upper Atmosphere," by A. V. Pokrovskii, director of the U. S. S. R. Institute of Experimental Aeromedicine.

The paper by Poloskov and Mirtov describes an instrument container, 6.5 feet long and 15.75 inches in diameter, used for upper-atmosphere research. It is essentially a metal cylinder divided into three sections. The lower section is hermetically sealed and contains power supplies, ammeters, camera, and the program mechanism which controls the operation of all the instruments in the container. The center section—which is open to the atmosphere—contains evacuated glass sampling flasks, thermal and ionization gages, etc. The upper section contains a parachute and is also hermetically sealed. A set of spikes in the bottom of the container ensures a vertical landing. The container, which weighs about 550 pounds, is jettisoned automatically in the descending phase of the trajectory at a height of 6 to 7.5 miles above the Earth's surface.

Pokrovskii's paper describes a catapultable chassis used in studying the behavior of dogs during round-trip flights to altitudes of 68 miles. The dog is secured in a hermetically sealed space suit with a removable plastic helmet and is provided with a 2-hour supply of oxygen. The chassis is equipped with radio transmitter, oscillograph, thermometers, pulse-measuring instruments, camera, and parachute. Two such chassis are fitted in the rocket nose section, which separates from the body of the rocket at the apex of the trajectory. One chassis separates from the nose section at a height of 50 to 56 miles and parachutes to the ground from a height of 46 to 53 miles. The other chassis separates at a height of 28 to 31 miles and falls freely to a height of 2 to 2.5 miles before parachuting to the ground.

As one might expect, the subject matter of these two papers received extremely wide publicity in the Soviet press. Probably the most comprehensive review was given by Academician Blagonravov himself in an article entitled "Investigation of the Upper Layers of the Atmosphere by Means of High-Altitude Rockets," which appeared in *Vestnik Akademii Nauk S. S. S. R.* in June 1957. Besides mentioning by name the key personnel in the program, Blagonravov stated that cosmic-ray investigations by means of rockets were initiated in the Soviet Union in 1947, that atmospheric composition studies to altitudes of 60 miles began in 1949, and that systematic studies of the atmosphere—including the use of dogs—were conducted from 1951 to 1956.

By way of interlude, a TASS dispatch datelined Moscow, June 18, 1957, reads as follows:

> At a press conference held by the State Committee for Cultural Relations with Foreign Countries on June 18, the correspondents were shown living travelers into extraterrestrial space —three dogs who were sent up in rockets to a height of 100 kilometers and more. Two of them have made two flights each and are in good health. All the flights were filmed. It was found

that the animals behaved normally when flying to this height at a speed of 1,170 meters per second. Alexei Pokrovskii, a member of the Soviet Committee for the International Geophysical Year, said, "I would like the British correspondents to inform the British Society of Happy Dogs about this because the society has protested to the Soviet Union against such experiments."

In June 1957, Academician I. P. Bardin submitted by letter to the CSAGI at Brussels the official U. S. S. R. rocket and Earth satellite program for the IGY. This program, which was merely an outline, indicated, among other things, that the Russians would fire 125 meteorological research rockets from three different geographical zones and would establish an unspecified number of artificial Earth satellites.

Of the numerous statements made by various Soviet scientists in the press and on the radio concerning the imminent launching of the first Soviet satellite, those by Academician A. N. Nesmeyanov were probably the most pertinent. On June 1, 1957, *Pravda* quoted Nesmeyanov as follows:

> As a result of many years of work by Soviet scientists and engineers to the present time, rockets and all the necessary equipment and apparatus have been created by means of which the problem of an artificial Earth satellite for scientific research purposes can be solved.

A week later, Nesmeyanov said that

> soon, literally within the next months, our planet Earth will acquire another satellite. . . . The technical difficulties that stood in the way of the solution of this grandiose task have been overcome by our scientists. The apparatus by means of which this extremely bold experiment can be realized has already been created.[6]

In addition to these guarded statements by Nesmeyanov, and those of other Soviet scientists, the scientific literature contains several specific indications of the forthcoming launching of the first Soviet satellite. For example, a one-page announcement entitled "On the Observation of the Artificial Satellite," by A. A. Mikhailov, Chairman of the Astronomical Council of the U. S. S. R. Academy of Sciences, appeared on page 1 in the astronomical journals *Astronomicheskii Tsirkulyar,* May 18, 1957, and *Astronomicheskii Zhurnal,* May-June 1957. After a brief description of what observers were to expect as the satellite passed overhead,[7] the announcement concluded with the following statements:

> The Astronomical Council of the U. S. S. R. Academy of Sciences requests all astronomical organizations, all astronomers of the Soviet Union, and all members of the All-Union Astronomical and Geodetic Society to participate actively in preparations for the visual observations of artificial satellites.
>
> Instructions and special apparatus for observation can be obtained through the Astronomical Council.[8]

Two articles in the June 1957 issue of the Russian amateur-radio magazine, *Radio*,[9] provide further evidence of the imminent establishment of Sputnik I. The articles, entitled "Artificial Earth Satellites—Information for Radio Amateurs," by V. Vakhnin, and "Observations of the Radio Signals from the Artificial Earth Satellite and Their Scientific Importance," by A. Kazantsev, gave a fairly comprehensive description not only of a satellite's orbit and how the subsequent appearances of a satellite can be predicted, but also of the satellite's radio transmitters, how the 20- and 40-megacycle frequency signals are to be used, and what information about the upper atmosphere can be derived from them.

The July and August issues of *Radio* carried articles on how to build a recommended short-wave-radio receiver and a direction-finding attachment for tracking the Soviet sputniks. Moreover, to inform the Russian radio amateurs about developments in the United States, the July issue of the magazine carried an article based on material taken from the American amateur-radio magazine *QST* describing the Minitrack II system which would permit radio amateurs to track American satellites with comparatively inexpensive equipment. This item was followed immediately by a notice in bold-face type to Soviet radio amateurs to make preparations for tracking the Russian scientific Earth satellites and contained detailed instructions on how to submit data on the signals received and recorded to Moskva—Sputnik for reduction and analysis by the Institute of Radio Engineering and Electronics of the U. S. S. R. Academy of Sciences.

That the Soviets were in earnest about their missile capabilities and space-flight intentions became indubitably clear on August 27, 1957, when a TASS report in *Pravda* stated that

> successful tests of an intercontinental ballistic rocket and also explosions of nuclear and thermonuclear weapons have been carried out in conformity with the plan of scientific research work in the U. S. S. R.

There was considerable prognostication that the Soviets would launch a satellite on September 17, 1957, the 100th anniversary of the birth of K. E. Tsiolkovskii, the founder of the science of astronautics. Needless to say, this day was the occasion for speeches by many leading scientists, both at the Hall of Columns in Moscow and at Peace Square in Kaluga, a small town approximately 100 miles southwest of Moscow, where Tsiolkovskii had spent the greater part of his life. At Kaluga the Soviets will erect a monument depicting Tsiolkovskii in flowing cape, looking into the sky, and standing on a pedestal in front of a long slender rocket poised in a vertical takeoff position.

The climax to this chronicle occurred, of course, on October 4, 1957, when Sputnik I was established in its orbit. Appropriately enough, even on this occasion the farsighted Soviets had scientific delegations strategically placed in foreign capitals. Washington played host to IGY delegates A. A. Blagonravov, V. V. Belousov, A. M. Kasatkin, and S. M. Poloskov, who were, naturally, overjoyed on hearing that the satellite had been launched successfully. In Barcelona, where the Eighth International Astronautical Congress was convening, a Soviet delegation of four, headed by L. I. Sedov, made the most of the occasion by presenting two papers, one by L. V. Kurnosova on the investigation of cosmic radiation by means of an artificial Earth satellite and the other by A. T. Masevich on the preparation for visual observation of artificial satellites, a Soviet version of the Moonwatch program. Sedov also distributed a limited number of copies of the special 284-page September 1957 issue of *Uspekhi Fizicheskikh Nauk,* which contains 17 papers on various aspects of Soviet rocket and satellite research.

B. Soviet Literature on Space Flight

Russia has a well-established literature on rocketry and space flight. This literature includes not only the classic works of her own pioneers, but also translations of foreign monographs by Esnault-Pelterie, Oberth, Hohmann, Goddard, Sänger, and others. The Soviets have also "liberated" a vast amount of detailed material from German industrial firms and scientific and technological institutes. Russian textbooks on rocketry, for instance, consider details of German developments that are not even mentioned in American books on the subject.[10]

In the post-Tsiolkovskii period the names of M. K. Tikhonravov and A. A. Shternfel'd stand out prominently in spite of the Stalin shadow. Both are capable and prolific writers. For several years the names of popular-science writers B. V. Lyapunov and M. V. Vasil'ev, engineers K. A. Gil'zin and Yu. S. Khlebtsevich, and scientist K. P. Stanyukovich have been appearing with increasing frequency on articles and books about rocketry and space flight. More recently Soviet scientists have been reporting the results of their researches not only in the technical journals but also in the popular press, either in the form of interviews or as nontechnical essays. Since 1951 a monthly journal, *Voprosy Raketnoi Tekhniki* (*Problems of Rocket Technology*), has been completely devoted to translations and surveys of the foreign periodical literature. Since 1954 the Institute of Scientific Information of the U. S. S. R. Academy of Sciences has been publishing a journal, *Referativnyi Zhurnal: Astronomiya i Geodesiya* (*Reference Journal: Astronomy and Geodesy*), which abstracts, among other things, foreign and domestic publications in the field of astronautics. Moreover, the Soviets have, of course, their own classified literature, which in all probability is extremely interesting.

Prior to 1955 Soviet papers on space flight followed, in general, a fixed pattern. They began with an account of the historical contributions made by the early Russian astronauts; next came a discussion of the test results obtained by American and other foreign rocketeers, followed by a disclosure of the problems involved in launching a satellite vehicle and of the variety and importance of the data to be obtained from an extraterrestrial laboratory; finally, they boasted about the great efforts that Soviet scientists were exerting in creating a scientific space station and in realizing cosmic flights possible for peaceful purposes. It is interesting to note that (except in one or two cases) almost no mention is made of any specific Soviet developments or results. Thus, for example, the article on rockets in the *Bol'shaya Sovetskaya Entsiklopediya*[11] includes two tables of rocket characteristics. Table 1 lists the characteristics of some liquid-fuel rockets, including the German A-4 (V-2) and the Wasserfall, the United States Viking No. 9 and the Nike, the French Véronique, and the United States two-stage Bumper (V-2 plus Wac Corporal) rocket. But no Russian rockets. Table 2 gives the characteristics of some rocket missiles, including the German Rheinbote and a 78-millimeter fragmentation shell, and the United States Mighty Mouse and Sparrow missiles. Again, no Russian missiles.[12]

In recent years Soviet papers on astronautics have become more and more specialized, dealing with such topics as chemical and nuclear rocket engines, radio guidance, meteoric impacts, weightlessness, and orbit calculations, as well as with problems to be investigated during the IGY. Their tone has been somewhat conciliatory to the West, and the jibes at the capitalist countries, ever present in the earlier papers, are conspicuously absent.

Articles on the problems of astronautics by topnotch Soviet scientists and technologists began to appear in the official, serious scientific publications of the U. S. S. R. Academy of Sciences in 1954. Typical of such articles are Shternfel'd's "Problems of Cosmic Flight," published in *Priroda* in December 1954, which is primarily an exposition on flight trajectories from the Earth to the Moon, Mars, Venus, and Mercury; Academician V. G. Fesenkov's "The Problems of Astronautics," which appeared in *Priroda* in June 1955 and which was written from the point of view of an astrophysicist who touches on the possibility of using atomic energy as a source of power for space travel; and "Contemporary Problems of Cosmic Flights," by A. G. Karpenko and G. A. Skuridin, published in *Vestnik Akademii Nauk S. S. S. R.* in September 1955. The last article is a comprehensive survey of the state of the art gleaned largely from papers presented at the Fifth and Sixth International Astronautical Congresses. It concludes, significantly, with the following statements:

> Together with the utilization of atomic energy for peaceful purposes and the development of the technology of semiconductors and new computing machines, the problem of interplanetary communications belongs with those problems which will open to mankind great areas of scientific cognition and the conquest of nature.

The importance of this problem was clearly described by Academician P. L. Kapitsa, a member of the Commission on Interplanetary Communications: ". . . if in any branch of knowledge the possibilities of penetrating a new, virgin field of investigation are opening, then it must be done without fail, because the history of science teaches that, as a rule, it is precisely this penetration of new fields that leads to the discovery of those very important phenomena of nature which most significantly widen the paths of the development of human culture. . . ."

The Russian technical literature of recent years gives abundant evidence of continued progress in the various disciplines associated with space flight. One of the most important pieces of evidence was the publication a few years ago of tables of thermodynamic properties, ranging from 298° K to 5,000° K, of such chemical species as F_2, HF, CH, CH_2, CH_3, and C_2. The first two indicate an interest in fluorine as an oxidant in chemical-rocket propellant systems, and the latter four, an interest in hydrocarbons as possible propellants in nuclear rockets.

The subject of nuclear-powered rockets is treated by K. P. Stanyukovich in an article entitled "Problems of Interplanetary Flights," which appeared in the August 10, 1954, issue of *Krasnaya Zvezda,* and in a slightly more expanded form as a paper entitled "Rockets for Interplanetary Flights" in the book *Problemy Ispol'zovaniya Atomnoi Energii* (*Problems of Utilizing Atomic Energy*), published in 1956. Diagrams of nuclear-powered turbojet, ramjet, and rocket engines illustrate G. Nesterenko's article, "The Atomic Airplane of the Future," published in *Kryl'ya Rodiny* in January 1956, while R. G. Perel'man's article, "Atomic Engines," in the January 1956 issue of *Nauka i Zhizn',* includes a sketch of a six-stage cosmic rocket in which the first stage is powered by a liquid-rocket engine, the second stage by a ramjet engine, the third stage by an atomic-rocket engine, and the three final stages by liquid-rocket engines.

In celebrating its 125th anniversary in 1955, the Moscow Higher Technical College, which is also known as the Bauman Institute and is the Russian counterpart of the Massachusetts Institute of Technology or California Institute of Technology, published a collection of 19 papers on theoretical mechanics, several of which had direct applications to space flight. One that is particularly relevant was written by V. F. Krotov and is entitled "Calculation of the Optimum Trajectory for the Transition of a Rocket to a Given Circular Trajectory around the Earth."

Rocket guidance has been discussed by a number of Russian experts, notably by I. Kucherov in an article entitled "Radio-guided Rockets," published in *Radio* in August 1955, and by Yu. S. Khlebtsevich, who wrote several articles on rocket flights to the Moon, Mars, and Venus.

Some highly interesting and original ideas have appeared in recent articles in Russian popular scientific literature. One article proposes worldwide television broadcasting by means of three Earth satellites symmetrically spaced in an equatorial orbit at an altitude of 22,200 miles. Needless to say, the author discreetly avoids mentioning the military significance of such a system. Another article suggests the use of Earth satellites for the experimental verification of the general theory of relativity. This article, written by V. L. Ginzburg, is a very lucid piece of nontechnical scientific writing on a subject that is generally considered too abstruse for the layman to understand. The study of the biological problems of interplanetary flight continues to be the subject of considerable discussion and investigation.

The two general subjects that have received the most attention in the Soviet press are the artificial Earth satellite and rocket flight to the Moon. Prior to Sputnik I's establishment in orbit, the following scientists wrote papers on the problems associated with artificial Earth satellites: K. P. Stanyukovich, "Artificial Earth Satellite," *Krasnaya Zvezda,* August 7, 1955; A. G. Karpenko, "Cosmic Laboratory," *Moskovskaya Pravda,* August 14, 1955; G. I. Pokrovskii, "Artificial Earth Satellite," *Izvestiya,* August 19, 1955; L. I. Sedov, "On Flights into Space," *Pravda,* September 26, 1955; and A. N. Nesmeyanov, "The Problem of Creating an Artificial Earth Satellite," *Pravda,* June 1, 1957. The first four articles were prompted largely by the White House announcement of July 29, 1955, while Nesmeyanov's article was a harbinger of Sputnik I.

In the Soviet literature there are repeated references to Moon-rocket projects. For example, in an article entitled "Flight to the Moon," published in *Pionerskaya Pravda* on October 2, 1951, M. K. Tikhonravov, corresponding member of the Academy of Artillery Sciences, stated that according to engineering calculations two men could fly around the Moon and back to Earth in a rocket ship weighing approximately 1,000 tons. Such a ship must have a velocity of approximately 6.9 miles per second. If an artificial Earth satellite were available, then it would be possible to send a much smaller space ship—one weighing not more than 100 tons and having a velocity of 2.2 miles per second—from the satellite to the Moon.

According to a German press agency report, the Soviet newspaper *Krasnii Flot* (*Red Fleet*) for October 12, 1951, asserted that a Moon rocket had already been designed in the Soviet Union. It was said to be 197 feet long, to have a maximum diameter of 40 feet, a weight of 1,000 tons, and 20 motors with a total power of 350 million horsepower. Heinz H. Kölle of Stuttgart's Gesellschaft für Weltraumforschung evaluated these data in an article entitled "Wird in der Sowjet-Union eine Mondrakete gebaut?" in *Weltraumfahrt,* January 1952. He concluded that in the optimum case a manned rocket for at best a two-man crew and a single circumnavigation of the Moon with subsequent return to the Earth still lay too close to the outermost limit of current attainments:

> Even the unmanned Moon messenger would require immense technical effort. The practical result would be trifling in comparison. On the other hand, the undertaking could be used psychologically and propagandawise, since successful execution and the corresponding accompanying fanfare would obviously demonstrate that Soviet long-range rockets would just as well reach any point on the Earth's surface.

Stanyukovich, a man of many interests and of prolific pen, has made several contributions in this field. His article "Trip to the Moon: Fantasy and Reality" in the English-language propaganda journal *News: A Soviet Review of World Events,* for June 1, 1954, is more polemic than scientific. His article "Rendezvous with Mars" in the same journal for October 16, 1956, is not quite so belligerent toward the United States. In this later article he predicts flights to the moon in 5 to 10 years and to Mars within 15 years, the latter being accomplished not with chemical fuels but with nuclear fuels.

Perhaps the most widely publicized Moon-rocket project in the Soviet Union is that proposed by Yu. S. Khlebtsevich, which made its first appearance in an article entitled "On the

Way to the Stars," in *Tekhnika-Molodezhi* in July 1954; later it was published in an expanded form as "The Road into the Cosmos," in the November 1955 issue of *Nauka i Zhizn'*. Khlebtsevich suggests landing a mobile "tankette-laboratory" on the Moon. The tankette, which would weigh not more than a few hundred pounds and would be radio controlled from the earth, would explore the surface of the Moon and report its findings back to earth.[13] Information so obtained would make possible the next stage—the mastery of the Moon by man in the next 5 to 10 years.

In February 1957, the Soviet press gave considerable publicity to a space-flight project headed by Prof. G. A. Chebotarev at the Institute of Theoretical Astronomy in Leningrad. According to Chebotarev's calculations it is possible, with the expenditure of only 16 tons of propellant, to launch a rocket vehicle weighing 110 to 220 pounds with an initial velocity of 6.8 miles per second in an elliptical orbit around the Moon. Flying solely under gravitational forces the vehicle would round the Moon at a distance of 18,600 miles and return to Earth in 236 hours, after covering a total path length of about 620,000 miles.

One of the most startling disclosures in connection with Soviet space-flight activities is the paper entitled "Some Questions on the Dynamics of Flight to the Moon" by V. A. Egorov of the Steklov Mathematics Institute in Moscow. This paper is a summary of a systematic investigation undertaken from 1953 to 1955 to find satisfactory solutions for the fundamental problems in the theory of flight to the Moon: specifically, the problem of the form and classification of unpowered trajectories, of the possibility of periodic circumflight of the Moon and the Earth, and of hitting the Moon. The paper also deals with the particularly important question of the effect of dispersion in initial data on the realization of hitting or circumflight. More than 600 trajectories were calculated by means of electronic computers and were classified as hits, circumflights, or afflights (that is, approach trajectories which do not encompass the Moon but allow one to see everything on its opposite side and to return to Earth). This investigation is quite similar to studies of the general trajectories of a body in the Earth-Moon system that are being conducted in this country. The overall results of the studies are in substantial agreement. Specific numerical comparisons can now be made, since the complete report is available.

C. The Soviet Ballistic-Missile and Space-Flight Program

The Soviets began to flex their ballistic muscles with the announcement on August 27, 1957, of a successful test in the Soviet Union of an intercontinental ballistic missile capable of carrying a powerful nuclear weapon to any point on the globe. The guidance system was said to be capable of placing the missile on target with an error not exceeding two thousandths of the range; i.e., for a flight range of 6,200 miles, the missile would not miss the target by more than 12.4 miles. G. I. Pokrovskii discusses the problem of attaining the precision required to put ballistic missiles on target in an article entitled "Architecture in the Cosmos" in the December 1957 issue of *Tekhnika-Molodezhi.* The last-stage engine accelerates the ballistic rocket to its assigned velocity in an "ethereal gun-barrel" or "tunnel" formed by radio beams from three or four radio stations on the ground. At the slightest deviation in direction the missile enters a zone in which the intensity of the radio waves is greater than in the center of the tunnel. The waves act on the missile's automatic-control instruments and return it to the center of the tunnel. A radio signal shuts off the last-stage engine at the precise moment at which the rocket has attained its predetermined speed.

To impress the world that their possession of the ICBM is fact, not fantasy, the Soviets followed through with an unprecedented display of propulsive might by launching, in quick succession, artificial Earth satellites on October 4 and November 3, 1957. The size of the sputnik carrier rockets was evident from the fact that, as K. P. Stanyukovich pointed out, they could be easily observed with the naked eye as stars of zero or first magnitude, whereas the American satellite Explorer I can be observed as a star of fifth or sixth magnitude only when it is closest to the earth. These differences in stellar magnitude indicate that the reflective areas of the sputnik carrier rockets were no less than 100 times greater than that of Explorer I. It is quite likely, therefore, that Sputnik II was no smaller dimensionally than the ballistic rockets displayed in Moscow during the Red Square parade on November 7, 1957, i.e., about the size of the United States Redstone missile. Moreover, since the announced weight of the experimental equipment in Sputnik II was 1,118 pounds, the entire device in orbit must have had a mass in the neighborhood of 4 to 7 tons.

In the March 1958 issue of *Astronautics,* Walter R. Dornberger (of V-2 fame) makes the following observation with regard to the Soviet missile and space-flight program: "Along with the experience they gained in handling long-range rockets, the Russians also got the Peenemünde way of thinking and the schedule for space conquest we had set up as far back as 1942. The satellites are only the first step. Another look at the schedule is all that's necessary to predict what lies ahead." The schedule that Dornberger and his confrères at Peenemünde had set up was the following ten-point guided-missile and space-flight program:

1. Automatic long-range single-stage rockets (A-4 or V-2).
2. Automatic long-range gliders (A-9B).
3. Manned long-range gliders (A-9B).
4. Automatic multistage rockets (A-9/10).
5. Manned hypersonic gliders (A-9B/10).
6. Unmanned satellites.
7. Manned ferry rockets to satellite orbits.
8. Manned satellites.
9. Automatic space vehicles.
10. Manned space vehicles.

Since the Soviets are masters in the arts of exploitation and long-range planning, as well as being endowed with a native competence in matters scientific and technological, it is not difficult to imagine the alacrity with which they assimilated the Peenemünde program and adapted it to their own plans for world domination. It is not known how slavishly they are adhering to the Peenemünde program, but it is known that they have already accomplished points 1, 4, and 6 and have made considerable progress in implementing some of the others.

The Soviet ballistic-missile and space-flight program is probably somewhat more involved and proliferated than the straightforward program of the Peenemünde group. It undoubtedly follows a logical pattern of development, involving the integration of complex military facilities and skills with the disciplines of the scientific and technical communities. The probable activities of the Soviet program can be arranged in four general categories that depend, in the main, on theoretical minimum-space-flight-velocity requirements and on the type of mission to be accomplished.

Category I is, in Russian parlance, that of geocosmic flights, or flights from one point of the terrestrial globe to another through cosmic space, for which the flight-velocity requirement is less than orbital (less than 4.9 miles per second). This category is the fundamental one, because the success of the remainder of the program depends on it. Soviet achievements in geocosmic flight consist of their long-range one- and two-stage (i.e., their much-heralded "multistage") ballistic missiles, as well as their geophysical rockets for exploring the upper atmosphere and their biological rockets for studying the behavior of animals during flight to and from the upper atmosphere. The Soviets have had considerable experience and success with their technique of recovering, by parachute, test-instrument containers and encapsulated experimental animals after rocket flights to altitudes exceeding 60 miles. In view of their extensive studies in the various aspects of space medicine and their patent desire to be first to achieve manned space flight, it is reasonable to assume that the Soviets will soon announce the "successful" return of a human passenger from a rocket flight in the Soviet Union.

Category II is that of orbital flight around the Earth, for which the flight-velocity requirement is between 4.9 miles per second (the so-called first cosmic or circular velocity) and 6.9 miles per second (the so-called second cosmic or "escape" velocity). In this category the Soviets have achieved three successful launchings that were (*a*) spectacular primarily because of the size of the packages placed in orbit; (*b*) significant because of their scientific and military reconnaissance implications; and (*c*) effective as propaganda devices.

Commenting on the level of developments of technology in the Soviet Union, Khrushchev made the following statement in a speech at Minsk on January 22, 1958: "The whole world was amazed by the fact that the second artificial satellite weighed over six times more than the first one; it weighed more than half a ton. But even this is not the limit. We can double, even more than double, the weight of the satellite, because the Soviet intercontinental rocket has enormous power which makes it possible for us to launch an even heavier satellite to a still greater height. And we shall probably do so!" On May 15, 1958, the Soviets announced that they had place Sputnik III in orbit. This satellite had a gross weight of 2,919 pounds, 2,130 pounds of which was instrumentation that was somewhat more sophisticated than that in Sputniks I and II.

A. V. Topchiev, chief scientific secretary of the U. S. S. R. Academy of Sciences, in summarizing the first scientific results obtained from Sputniks I and II before a general assembly of the academy in March 1958, formalized Khrushchev's statement in the following terms:

> Using the achievements of native reactive technology which make it possible to raise to a great height and place in orbit containers with scientific apparatus weighing many hundreds of kilograms, our scientists can now raise the most diverse problems in the investigation of the upper layers of the atmosphere and in the region of cosmic space closest to the Earth. It is clear, also, that the solution of the problem of long flights in cosmic space and the attainment of other planets lies only in creating satellites of great weight.

The Soviets are quite aware of the importance of recovering photographic film, instruments, and animals from satellites; the examination of such items after orbiting around the Earth would be of much greater value than telemetered data. They have discussed the general techniques of recovery and have indicated the existence of various recovery projects. One such project, described in some detail by V. Petrov,[14] involves the ejection of a 20.9-pound package from a satellite or carrier rocket at minimum orbital altitude. The package consists of an 11.9-pound retrorocket and a 3.5-pound stainless-steel collapsible sphere, which, when inflated with helium, measures 3 feet in diameter and acts as a drag brake, bringing down to Earth, with a terminal velocity of 29.5 feet per second, a small beacon transmitter and an 8-ounce cartridge of exposed film within 20 minutes after leaving the orbit. Prof. E. K. Federov, commenting on the prospects of recovering satellites, said that the problem is solvable in principle, but as yet has not been solved. At any rate, the availability of such massive sputniks should afford Soviet scientists and technologists abundant opportunities for testing their recovery techniques.

The ultimate goal in this category is, of course, a manned space station that can serve not only as a space laboratory, but also as an intermediate station for future interplanetary voyages. The realization of this goal presupposes the implementation of points 3, 5, and 7 of the Peenemünde program.

Category III is that of lunar flight, for which the minimum flight-velocity requirement is about 6.8 miles per second—slightly less than escape velocity. The Soviets are admittedly deeply engaged in a Moon-rocket program. Professor Stanyukovich has indicated that "if a few more stages were added to modern ballistic rockets, then the last stage of such a rocket would attain a speed of 12 kilometers per second (7.4 miles per second). This will be quite sufficient to fly to the Moon. The first flight to the Moon, or circumflight of the Moon, will evidently take place within the next few years." But, he added, "before a rocket flies to the Moon, a number of artificial satellites will be launched along increasingly elongated elliptical orbits which will draw nearer and nearer to the Moon. Instruments installed in such satellites will make it possible to closely study and to photograph the lunar surface, and to learn the nature of its mysterious relief." [15] Considering the extensive calculations of Earth-Moon trajectories that have been carried out by V. A. Egorov at the Steklov Mathematics Institute in Moscow and by G. A. Chebotarev at the Institute of Theoretical Astronomy in Leningrad, the Soviets are theoretically well prepared for lunar flights. There seems to be no question of their propulsion capability in this respect, but it remains to be seen whether they have the necessary guidance and control capability to strike the Moon. Soviet propulsion capability was demonstrated on January 2, 1959, with the launching of the lunar rocket Mechta. In a sense this rocket accomplished the mission that the Army lunar rocket Pioneer II was intended to accomplish (i.e., either to strike the Moon or, failing that, to orbit around the Sun).

Category IV is that of interplanetary travel, for which the flight-velocity requirement is considerably greater than escape velocity because of the maneuvers involved in transferring from one planetary orbit to another. Thus, for a spaceship to fly with a minimum expenditure of propellants from Earth to Mars along an ellipse tangent to the orbits of both planets, the theoretical minimum velocity required for a hard landing (impact) on Mars would be 10.4 miles per second, and for a soft landing, about 13.5 miles per second. The corresponding theoretical minimum velocities for an Earth-to-Venus flight would be 10.2 and 16.6 miles per second, respectively. The Soviets, aware of the limitations of chemical rockets in this regime, are assertedly looking forward to the role that nuclear engines will play in the future—not only in interplanetary flights, but in geocosmic flights as well.

In view of the variability of interplanetary distances, flights from Earth to other planets will be scheduled, not arbitrarily, but in accordance with a rigid timetable based on the most favorable conjunction of the planets with Earth. It is entirely possible that the Soviets, after a successful lunar impact, might attempt to send rockets to Mars and/or Venus along tangential orbits as mentioned above. The duration of these excursions to Mars and Venus would be 260 and 146 days, respectively. The probable launching dates can be determined by reference to "Specific Flight Possibilities," in chapter 20.

Because the space-flight program is inherently connected with the ICBM, Soviet reluctance to discuss certain details of satellite launching is necessarily dictated by military secrecy. It is ironic, however, that the more sputniks that are placed in orbit, the more the free world will learn about Russian military capabilities in rocketry as a result of direct observation and logical deduction. The following two examples can be cited.

Prof. Tadao Takenounchi, of the University of Tokyo Astronomical Laboratory, on the basis of sputnik periods and initial transit times over Moscow, published by TASS, determined the launching time of Sputnik I to be 1921 hours (GMT) October 4, 1957, and that of Sputnik II to be 0232 hours (GMT) November 3, 1957. The intersection of the traces of the two satellites, the elements of whose orbits were determined from observations made in Japan, placed the launching site in the Kyzyl Kum Desert at a spot with the approximate coordinates 42°30″ N./65°00″ E., i.e., about 248 miles southeast of the Aral Sea. Data from Sputnik III should help to establish the launching site somewhat more precisely.

In August 1958, at the International Astronautical Federation Congress in Amsterdam, Academician L. I. Sedov, Chairman of the Soviet Union's Commission on Astronautics, presented a paper on "Dynamic Effects in the Motion of Artificial Earth Satellites." For illustrative purposes, he presented a table of parameters of the orbits at the beginning of motion of each of the three Soviet scientific satellites. Of special interest is the fact that the values of perigee altitude (i.e., the minimum altitude) were 140-141.3, 139.5, and 140 miles, for Sputniks I, II, and III, respectively. The corresponding values of apogee altitude (i.e., the maximum altitude) were 587, 1,036, and 1,166 miles, respectively. The almost identical values of perigee altitude may indicate that a fairly good guidance system was employed by the Soviet scientists in placing their satellites into orbit, although not necessarily one suitable for ICBM use.

Because the Astronomical Council of the U. S. S. R. Academy of Sciences, through its permanent ICIC, is allegedly the agency responsible for the conduct of the Soviet space-flight program (at least the scientific aspects thereof), it is logical to expect that certain phases of this program would come to light in the pages of the Academy's scientific journals. This has, indeed, been the case. Even before Sputnik I was launched, a number of space-flight research papers appeared in a variety of journals. In celebration of the 100th anniversary of the birth of K. E. Tsiolkovskii, the Academy saw fit to publish 17 papers pertaining to astronautics in the September 1957 issue of *Uspekhi Fizicheskikh Nauk* (*Advances in the Physical Sciences*). Since the sputniks, the U. S. S. R. Academy of Sciences has been publishing the results of its theoretical and experimental space-flight research, not only in its established scientific periodicals, but also in special collections such as the one entitled "Preliminary Results of Scientific Researches on the First Soviet Artificial Earth Satellites and Rockets," dated July 1958. It is of interest to note that during the August 1958 IGY meeting in Moscow at the rocket and satellite symposium, Soviet scientists presented 18 papers based on research conducted by means of rockets and satellites. No information was released concerning the nature of the rockets used to put the Soviet satellites into orbit.

In reading Soviet scientific literature, one cannot help but be impressed by the boldness, the scope, and the dedication of the Soviet effort toward the ultimate conquest of the cosmos; i.e., manned interplanetary travel. There seems to be no question in the Soviet mind that the Communist Party's authoritarian directive to realize man's most cherished dream will be fulfilled in and by the Soviet Union. The Soviets have already made great strides toward attaining this goal by virtue of their large geophysical research rockets and by their massive artificial Earth satellites and extensive detection, tracking, and data-handling network associated with the Earth-satellite program. In addition to these established and well-publicized advances, the Soviets are building up a tremendous backlog of detailed information from concentrated studies in geophysics, astrophysics, celestial mechanics, radio astronomy, planetology, astrobiology, space medicine, and a variety of other disciplines, an intimate knowledge of which will help make interplanetary communication a reality.

Soviet confidence in the ultimate result is reflected in the fact that, whereas originally authorities exhorted sputnik observers—professional as well as amateur—to send their tracking data to Moskva–Sputnik, they now ask that data be sent to Moskva–Kosmos.

Notes

[1] E. g., Proceedings of the All-Union Conference on the Study of the Stratosphere, March 31-April 6, 1934, U. S. S. R. Academy of Sciences, 1935, and collections of papers titled "Rocket Technology" and "Jet Propulsion," Union of Scientific Technical Publishing Houses, 1935, 1936. Unfortunately, few, if any, of the latter items reached the United States.

[2] A similar development in this country led to the 75,000-pound-thrust rocket engine for the Redstone missile.

[3] Pravda, November 28, 1953.

[4] In Russian the term "interplanetary communications" is synonymous with "astronautics" and "space flight."

[5] Dr. V. A. Troitskaya, scientific secretary of the Soviet National Committee, read the accompanying English version immediately after Bardin's original statement.

[6] Komsomolskaya Pravda, June 9, 1957.

[7] Readers will find a striking similarity between this description and that of the Moonwatch program of the Smithsonian Astrophysical Observatory at Cambridge, Mass., the various aspects of which are described in the Observatory's Bulletin for Visual Observers of Satellites which began publication in July 1956. This

Bulletin, issued at irregular intervals, may be found as a center insert in the monthly journal Sky and Telescope.

[8] The telescopes used by members of Russian Moonwatch teams, as shown in photographs in Pravda and other Russian newspapers, after the launching of Sputnik I, are suspiciously similar in outward appearance to the design described in the Bulletin for Visual Observers of Satellites.

[9] Radio is an organ of the U. S. S. R. Ministry of Communication and of DOSAAF (the All-Union Volunteer Society for the Promotion of the Army, Aviation, and Navy) and corresponds to the American amateur-radio magazine QST, published by the American Radio Relay League.

[10] See, for example, Bolgarskii and Shchukin, Rabochie protsessy v zhidkostno-reaktivnykh dvigatelyakh (Working processes in liquid-jet engines), Oborongiz, Moscow, 1953, 424 pages.

[11] 2d edition, vol. 35, pp. 665-668.

[12] There is, however, a comprehensive table of Soviet missiles and their characteristics, prepared by Alfred J. Zaehringer, in the Journal of Space Flight, May 1956.

[13] This project has been made the subject of a Russian popular-science short film—of the Walt Disney type, but much inferior—and is No. 15 in a series generally entitled "Science and Technology." Since the advent of Sputnik I, the film has been shown in movie theaters throughout the United States.

[14] Leningradskaya Pravda, November 17, 1957.

[15] Sovetskaya Aviatsiya, January 1, 1958.

29

Astronautics in Other Countries

A. Astronautics in the United Kingdom

British interest and activities in space flight and the requirements for its realization were for a long time identified with the British Interplanetary Society, a respected organization founded in 1933, that has regularly published a responsible and sober journal. The society has recently suggested that the resources of the British Commonwealth be pooled in the pursuit of a British Commonwealth Space Exploration Program.[1]

In addition to this essentially private space activity, the British Government has carried on development work of direct value to an actual astronautics capability. The most prominent pieces of work are the Black Knight and Blue Streak missile programs.

The Black Knight, a single-stage, liquid-propellant rocket vehicle, 35 feet in length, was fired on September 7, 1958, at the Woomera guided missiles test range in Australia.[2, 3] This test range is reported to be 1,200 miles long.[4] In this first test, Black Knight reached a maximum speed of 8,000 miles per hour and an altitude of 300 to 400 miles.[5, 6]

It is reported that Black Knight can reach an altitude of 600 miles, and that an effort will be made over the next year to combine Black Knight with a solid-propellant second stage to reach an altitude of 1,600 miles.[7]

The Blue Streak, now in development, is a ballistic missile to be stored and fired from underground, with a range reported to be well over 2,000 miles.[8] A combination of Blue Streak and Black Knight could launch a satellite weighing as much as 1,000 pounds.[9] Such rocket performance also implies a capability to launch instrumented probes on lunar and interplanetary flights.

The British Government has been considering, together with the Royal Society, what further steps should be taken toward a satellite program. This investigation is proceeding under the general premise that any British contribution should be unique and original, rather than something that may repeat what others have done.[10]

B. Astronautics in the People's Republic of China

There have been a number of reports of various kinds that claim an active space program in Red China. [11-14] According to these reports, an effort is going forward to launch a satellite, or even a Moon rocket, from the territory of Communist China, probably with Russian equipment. There is no information about the nationality of the personnel involved.

The objectives behind any such work, if it is actually taking place, are matters of speculation, but the prestige aspects, as related to Red China's desires for international recognition, would probably be a large factor.

The Red Chinese are apparently cooperating with the U. S. S. R. in the program of observation of Soviet satellites.[15]

Notes

[1] Industry Report, Missile Design and Development, vol. 4, No. 10, October 1958, p. 12.

[2] Black Knight Rocket Fired at Woomera, London Times, September 8, 1958, p. 8.

[3] Black Knight Re-entry Successful in First Australian Firing Effort, Aviation Week, vol. 69, No. 11, September 15, 1958, p. 33.

[4] Space Research Minus Controls Suggested, Los Angeles Times, November 10, 1958, pt. III, p. 1.

[5] Black Knight Test Rocket Designed for ICBM Nose Cone Evaluation, Aviation Week, vol. 69, No. 12, September 22, 1958, p. 50.

[6] International Cooperation in the Exploration of Space, staff report of the Select Committee on Astronautics and Space Exploration, 85th Cong., 2d sess., October 15, 1958.

[7] See footnotes 3 and 5.

[8] See footnote 2.

[9] See footnote 6.

[10] See footnotes 1, 3, and 6.

[11] Periscoping the World, Newsweek, vol. LII, No. 3, July 21, 1958, p. 12.

[12] What's News, The Wall Street Journal, September 30, 1958, p. 1.

[13] Red Chinese Space Projects Going Forward, Los Angeles Times, November 4, 1958, pt. I, p. 11.

[14] From One Who Fled—A Look at Science in Red China, U. S. News & World Report, vol. XLV, No. 19, November 7, 1958, p. 107.

[15] Astronautics and Space Exploration, hearings before the Select Committee on Astronautics and Space Exploration, 85th Cong., 2d sess., April 15 through May 12, 1958, on H. R. 11881; L. V. Berkner, p. 1043.

APPENDIX

CORRESPONDENCE BETWEEN THE SELECT COMMITTEE ON ASTRONAUTICS AND SPACE EXPLORATION AND THE RAND CORPORATION

SELECT COMMITTEE ON ASTRONAUTICS
AND SPACE EXPLORATION,
HOUSE OF REPRESENTATIVES,
Washington, D. C., November 14, 1958.

Mr. F. R. COLLBOHM,
President, The RAND Corporation,
Santa Monica, Calif.

DEAR MR. COLLBOHM: The Select Committee on Astronautics and Space Exploration of the House of Representatives is completing its work before turning over its responsibilities to the new standing Committee on Science and Astronautics in January 1959. Our major accomplishment has been the writing of the National Astronautics and Space Act of 1958, which we hope will be an important step toward assuring that our country shows the necessary leadership in the field of the space sciences and their practical application to strengthen the position of the United States in the larger struggle.

Our committee is very conscious that the job is not complete, even so far as the Congress is concerned, although the executive branch has been supplied with some of the tools for pushing space development. There will be future decisions on directions, speed, priorities, and many other matters which will come before the Congress under our system of government.

We are particularly conscious that a major responsibility of this committee is to aid the Congress in gaining an understanding of the new technologies and their implications for future defense, economic application, and scientific advance, and what some of the less tangible although very real psychological and political potentialities may be.

In light of these responsibilities, it is important that this committee prepare for the incoming Congress and the successor committee the best possible appreciation of the state of the art and the future trends which are likely to result from aggressive development of our capabilities, or conversely, from the failure to do so. The committee staff has given considerable thought to these matters and for a year has diligently sought to learn the opinions of those best qualified and to evaluate these matters for the committee. But they and I recognize that the final report on so grave a matter must be the most authoritative assessment which it is possible to give the American people. We have had many offers of help from different organizations, and their advice has been beneficial. However, we have come to the conclusion after careful review, that The RAND Corporation could make a unique contribution to the cause of public understanding, if we can persuade it to marshal its resources of long experience and talent in this field to help us prepare a balanced report on the space outlook suitable for public release. We particularly like RAND's reputation for independence and integrity.

I appreciate the importance to RAND as an independent, nonprofit corporation of confining its work to technical comments and scientific analysis and conclusions, since expressions of opinion on policy or administrative matters are the purview of the Congress or the executive as responsible branches of the Government. I am in complete agreement with this view. The committee wishes assistance on those matters in which RAND has scientific competence where we think you can contribute to a better understanding of the implications of the scientific possibilities in the field of space technology.

I have said nothing about the form any report prepared by your company might take, whether it would be printed as received with direct credit or whether it would serve better as grist for the mill for a report issued under committee staff auspices. I would prefer to leave this question to the good judgment of your people and mine when there is something concrete to consider. You may rest assured that we would not tamper with the content of a RAND report issued by us without full consultation and consent by your people. On the other hand, it is clear that our public responsibilities require that we review any material before printing it with our sponsorship. I do not anticipate any difficulties, of course, or I would not have made such a request for assistance of you.

There is a classified report which I will not describe specifically in this letter which your organization prepared for us earlier in the year. It was an excellent job, and I hope that in expurgated form it can be at least a point of departure for any new study for us.

I hope that with this explanation of our needs, I can have your agreement in principle with a plan for RAND cooperation with our committee. If you concur, I believe your people and mine on the committee staff who have already been in consultation are prepared to go ahead with specific arrangements.

The committee and I personally will be very grateful to you for such help as you can render.

Very sincerely,

JOHN W. MCCORMACK,
Majority Leader and Chairman.

THE RAND CORPORATION,
Santa Monica, Calif., December 1, 1958.

HON. JOHN W. MCCORMACK,
Chairman, Select Committee on Astronautics and Space Exploration, House of Representatives, Washington, D. C.

DEAR MR. CHAIRMAN: Thank you for your letter of November 14 in which you request that The RAND Corporation assist the House of Representatives Select Committee on Astronautics and Space Exploration in the preparation of its final report. We are honored by your solicitation of our technical comments based on our scientific analyses in the field of astronautics, which as you know have been and continue to be sponsored largely by the United States Air Force under the Project RAND contract. Also, we deeply appreciate your generous remarks on RAND's reputation for independence and integrity.

We welcome this opportunity to be of service to your committee as outlined in your letter to us. In accordance with RAND's purposes as a nonprofit organization to further and promote public welfare and national security, we concluded it to be appropriate that work by our staff in response to your request be supported with RAND Corporation funds.

We are aware of the importance of the committee's responsibilities to the Congress, to the successor Committee on Science and Astronautics, and to the public. Consequently, we have been proceeding as rapidly as possible in the collection and preparation of relevant material ever since the informal inquiry from the committee staff. We plan to have a preliminary draft of our response available to your committee this week. We certainly understand your requirements to review any material prior to printing and stand ready to consult with members of your committee and staff on the content of material furnished by RAND.

We trust that our contribution will assist the committee to build public understanding of the state of the art as well as of the uncertainties and complexities associated with the development and accomplishment of a vigorous, adequate astronautics and space exploration program.

Very sincerely,

F. R. COLLBOHM, President.

GLOSSARY

ablation cooling. One means of protecting a payload from the high external heating associated with vehicle re-entry into the atmosphere. The vehicle shell is designed to carry away heat by vaporizing or sloughing away (ablating) during re-entry.

absolute zero. The lowest temperature that can be reached, —273.16° Centigrade or 0° Kelvin.

absorptivity. The capacity of a surface for absorbing thermal radiation (heat), expressed as the ratio of absorbed to incident radiation. A good absorbing surface is a good radiator but a poor reflector. (See *emissivity.*)

accelerometer. An instrument for measuring the accelerations experienced by a vehicle in flight.

afflight. An approach trajectory that does not encompass the Moon but allows the vehicle to "see" everything on its opposite side before returning to Earth.

albedo. The reflecting power of an object (planet or satellite), expressed as the ratio of light reflected from an object to the total amount falling on it.

aphelion. That point in the orbit of a planet of the solar system where the planet is farthest from the Sun. (See *perihelion.*)

apogee. That point in the orbit of the Moon or of an Earth satellite where the object is farthest from the Earth. (See *perigee.*)

ascent trajectory. The elliptical path described by a vehicle from point of launch to maximum altitude (*apogee*).

astronomical unit. A unit of measure used for expressing distances in the solar system. One astronomical unit (a.u.) is the *mean distance* of the Earth from the Sun, about 92,897,000 miles.

ballistic missile. A vehicle that is guided or controlled during the initial flight phase only, its flight during the later phases assuming the characteristics of a freely falling body.

beacon. A device containing an automatic radar receiver and transmitter. Also called a transponder, it is used in guidance and warning systems.

binding energy, neutron. The energy required in removing one neutron from a nucleus, measured in millions of electron volts.

bipropellant. A liquid propellant consisting of two unmixed chemicals (fuel and *oxidizer*) that are kept separate until injected into the combustion chamber.

booster. A rocket used to increase or augment the speed, range, or altitude of the vehicle to which it is attached; the first *stage* of a multistage missile.

boundary layer. The thin, nearly motionless layer of air at the surface of a vehicle flying through the atmosphere. The internal mechanics of the boundary layer are of great interest in aerodynamic studies.

Bremsstrahlung. A German word meaning "braking or impact radiation" and denoting, in atomic physics, X-ray formation due to the deflection of one charged particle by another.

cinetheodolite. An instrument, combining a camera and telescope, used in *tracking* to determine and record the position-time history of an object in flight. The photographic record indicates time, azimuth and elevation, and position of the object with respect to cross-hairs in the telescope.

circumflight. Orbital flight around the Earth, Moon, or other celestial body.

circumlunar. Pertaining to orbital flight around the Moon.

composite. One of two general types of solid propellant. In a composite propellant the separate fuel and *oxidizer* are intimately mixed into one solid *grain.* (See *double-base.*)

continuous wave. An uninterrupted flow of radio energy. In tracking, the continuous wave from a radar is reflected by the target, but because of the relative motion of the tracker and target the frequency of the reflected wave keeps changing (*doppler effect* or doppler shift). Comparison of the transmitted and received frequencies enables measurement of range.

conversion efficiency. A fraction representing the efficiency with which a vehicle's kinetic and potential energy is converted into heat energy and transferred to the vehicle surface.

cooled detector. A device that indicates the presence of radiant energy (in radio, a demodulator that extracts the information from the modulated carrier wave), cooled to reduce internal noise. A cooled detector is thus sensitive to weak signals.

Coriolis force. A deflecting force due to the Earth's rotation, diverting moving objects to the right in the northern hemisphere, and to the left in the southern hemisphere. A man moving about in a rotating space vehicle would experience Coriolis effects.

descent range. In recovering a satellite payload, the ground distance covered by the payload from point of ejection to point of impact.

detection. The process of locating an object (in space), usually with optical, infrared, or radar and radio systems.

doppler effect. The apparent change in frequency of sound, light, and radar waves as the source and the observer move toward or away from each other. This effect is illustrated by the common experience of hearing the sound of an approaching train whistle suddenly drop in pitch as the train passes by.

double-base. One of two general types of solid propellant. A double-base propellant consists of nitrocellulose and nitroglycerine, plus stabilizers. (See *composite.*)

double star. A system of two stars related by mutual gravitational attraction and revolving about their common center of mass.

drain rate. The rate at which electrical power in a system is consumed.

Dyna-Soar. A pilot-controlled, military space vehicle designed to circle the Earth at orbital velocity, with controlled aircraft landing capabilities. (See section "Dyna-Soar.")

eccentricity. A numerical relation defining the shape of an ellipse. Taking a circular orbit to have an eccentricity of zero, elliptical orbits would have values approaching 1.00 as the orbit becomes more elongated.

ecliptic. The plane of the Earth's orbit, which makes an angle with the equator of about 23°27′.

ejection velocity. The velocity required to divert an object (payload) from a satellite orbit onto a trajectory that intersects the Earth's surface.

emissivity. The capacity of a surface for emitting thermal radiation (heat), expressed as the ratio of emissive power to that of a so-called blackbody (the complete radiator) at the same temperature. (See *absorptivity.*)

equatorial orbit. An orbit in the plane of the Earth's equator. A satellite in such an orbit would, in theory, be directly over the Earth's equator at all times.

escape velocity. That velocity required at a given location for a space vehicle to establish a parabolic (open-ended) orbit, thus escaping from a planet's gravitational field. The escape velocity for an Earth-launched vehicle is approximately 6.9 miles per second. (See *orbital velocity.*)

exosphere. The outermost fringe or layer of the atmosphere; the other layers, moving out from the Earth, are the troposphere, stratosphere, and *ionosphere.*

filaments. Fibered materials, or "whiskers," the use of which in structural elements promises high strength and low weight.

g. A symbol for the acceleration due to gravity at the surface of the Earth (approximately 32 feet per second each second). A man at rest on Earth is constantly subjected to 1 g, which translates into a force equal to his own weight in pounds; hence, a man accelerated at 10 g's would be subjected to a force ten times his own weight.

geocosmic flights. Flights from one point of the *terrestrial* globe to another through cosmic space, for which the flight-velocity requirement is less than orbital (less than 4.9 miles per second).

gimbals. A type of mounting that permits an object, such as a telescope or gyroscope, to turn freely in any direction.

grain. A shaped chemical mass of solid propellant, formed either by casting or by extrusion.

hard landing. An uncontrolled impact landing arbitrarily defined as one in which the approach velocity is greater than 500 feet per second.

high-energy propellants. Those that give a high *specific impulse,* such as fluorine and hydrazine.

hydyne. A proprietary name for a rocket fuel mixture.

hypergolic. Pertains to self-igniting propellants. The fuel and *oxidizer* ignite spontaneously on contact.

image intensifier. A device used with an *image orthicon* for amplifying photoelectrons and, by virtue of its selectivity, increasing the contrast between an object being tracked and the sky background.

image orthicon. A form of television camera tube used in *tracking,* in which an electron image is produced and then scanned by an electron beam.

inclination. That element of an orbit which indicates the angle between the plane of that orbit and a reference plane (in the solar system, the *ecliptic*). (See *orbit elements.*)

infrared. The range of electromagnetic radiation frequencies just above the visible spectrum. Detection systems sensitive to infrared have wide military application, since all bodies above absolute zero give off infrared radiation.

initial conditions. A set of parameters in effect at the beginning of a calculation, operation, test, etc.

interferometer. An instrument that separates a beam of light into two or more parts and causes them to travel different paths to a given point. Comparison of the phase difference of reunited beams that have traveled unequal paths permits very precise measurements. The phase-comparison method is used in some tracking systems, where the radio waves transmitted by a missile are received out of phase by separate ground antennas; i.e., the radio waves do not reach their maximum and minimum values simultaneously at each antenna. Measurement of the phase difference helps to locate the missile.

ionosphere. The third layer of the atmosphere. (See *exosphere.*)

ion rocket. A rocket whose propellant (usually an alkali metal) is ionized (each particle becomes an ion, acquiring an electrical charge), the ions then being accelerated through a nozzle by an electrical field to achieve a very high specific impulse.

isotopes. Atoms having the same atomic number but differing in atomic weight and radioactive properties. An element may have both stable and radioactive isotopes.

lift-drag ratio. In aerodynamics, the upward force (lift) on a vehicle due to airflow over a lifting surface, divided by the resistance (drag) of the vehicle to motion through the air.

light-year. The distance traveled by light in a year; approximately 6,000,000,000,000 miles.

linear accelerator. A device for producing high-energy ions or electrons by accelerating them along a straight path.

LOX. Jargon for liquid oxygen.

maser. A type of microwave amplifier capable of amplifying very weak signals without interference from internally generated noise.

magnetohydrodynamics. The study of the behavior of ionized gases (gases whose particles have acquired an electrical charge) when acted upon by electric and magnetic fields.

mean distance. Average distance. In astronomy, the mean distance of a planet from the Sun is the average of the *perihelion* and *aphelion* distances.

Microlock. The *tracking* and communication system for the Explorer satellite. The tracking method used is based on the principle of the *interferometer.*

mil. One ten-thousandth of an inch.

Minitrack. The *tracking* and communication system for the Vanguard satellite. The tracking method used is based on the principle of the *interferometer.*

monopropellant. An unstable liquid propellant that, under proper conditions, decomposes to form its own *oxidizer* and fuel, e.g., hydrogen peroxide.

nautical mile. Formerly 6080.20 feet in the U.S. On July 1, 1954, the international nautical mile of 6076.1033 feet was officially adopted in the U.S.

orbit elements. The six quantities that must be calculated to define the orbit of an object at any particular instant and to predict its position in the orbit at any future time. One such element is the angle between the orbit plane and the *ecliptic.*

orbital velocity. That velocity required at a given location for a space vehicle to establish an elliptical (closed) orbit around a planet. The orbital velocity for an Earth-launched vehicle is between 4.9 and 6.9 miles per second. (See *escape velocity.*)

overpressure. That pressure, caused by the explosion of a nuclear device, in excess of normal atmospheric pressure.

oxidizer or *oxidant.* That component of a propellant combination which provides the oxygen for combustion, e.g., liquid oxygen.

parameter. An arbitrary constant, as distinguished from an absolute constant. A parameter has a particular value in a given case, e.g., missile gross weight, whereas an absolute constant, such as π, always has the same value.

path angle. The angle between the velocity vector (direction) and the instantaneous horizontal.

Peenemünde. A German research station, constructed in 1937, on the Baltic coast. This was the center of German missile activity before and during World War II. Peenemünde is perhaps best known as the birthplace of the V-2 rocket.

perigee. That point in the orbit of the Moon or of an Earth satellite where the object is nearest the Earth. (See *apogee.*)

perihelion. That point in the orbit of a planet of the solar system where the planet is nearest the Sun. (See *aphelion.*)

period. The time required to complete one cycle or revolution. An Earth-period satellite would complete one revolution about the Earth in exactly 24 hours.

photon rocket. A rocket whose thrust is achieved by an internally generated beam of photons (quanta or "bundles" of radiant energy, such as light) emitted in a focused beam.

plasma rocket. A rocket whose gaseous propellant is partially ionized (the gas particles become electrically charged) by the discharge of a powerful arc through it, resulting in very high temperatures and high specific impulse.

point mass. For the purposes of calculation, the center of an object (planet), at which point the mass and forces of attraction are concentrated.

pulse. Radio energy transmitted by a pulse radar system for a very brief period, usually millionths of a second. In tracking, measurement of the time between transmission of a pulse and reception of an echo pulse from the target establishes the range to the target.

radiation pressure. The very feeble pressure exerted by electromagnetic radiation upon any surface exposed to it. (See *solar sail.*)

radio astronomy. The study of celestial bodies with the aid of radio waves (radio telescope) rather than with a conventional (optical) telescope.

radio-ranging. Pertaining to the transmission of radio beams as directional signals.

recovery. Landing a missile or payload safely at the end of flight. Some of the devices used to reduce landing impact are frangible structure, landing spikes, parachutes, and *retrorockets.* (See section "Landing and Recovery.")

relative biological effectiveness (RBE). The degree of biologic change produced in tissue by a dose of a certain type of radiation as compared with the corresponding change produced by an equal dose of a reference radiation (conventionally, X- or gamma radiation).

retrorocket. An accessory rocket used to decelerate a payload or to separate a burned-out *stage.*

roentgen-equivalent-physical (*r.e.p.*). A measure of radiation energy absorbed by tissue; specifically, 1 r.e.p. equals 93 ergs absorbed per gram of tissue.

soft landing. A controlled impact landing arbitrarily defined as one in which the approach velocity is less than 500 feet per second. (See *hard landing.*)

solar sail. Any large, lightweight surface designed to receive thrust from the *radiation pressure* of solar rays.

sounding rocket. A meteorological rocket used to gather data on (to "sound") the atmosphere at various altitudes.

specific impulse. A measure of the energy available from a propellant, being the thrust developed by burning 1 pound of fuel in 1 second.

sphereographical navigation. A variation on the theme of celestial navigation, but based on two successive observations (visual or radio) of a satellite instead of a star.

stage. In a missile composed of several sections, those independent sections which contain a powerplant, and which are progressively fired and jettisoned ("staged").

synoptic meteorology. That branch of meteorological science dealing with day-to-day variation of world-wide weather.

terrestrial. Pertaining to the Earth, as opposed to celestial.

test stand. Generally a fixture for holding a jet engine or rocket motor during a runup or firing for test purposes.

thermocouple. A device composed of two different metals that produces a current when heated.

tracking. The act of observing the progress of an object in flight. Because of the distances involved, tracking usually requires the use of optical or radar systems.

transpiration cooling. The pumping of gas or vapor through a porous skin to carry heat away from a vehicle as a means of protecting against re-entry heating.

transponder. See *beacon.*

Van de Graaff machine. An electrostatic generator used to accelerate electrically charged particles to energies high enough to cause nuclear reactions.

velocity of approach. The velocity of a payload just before contact with the "target" material.

weightlessness. A condition characterized by absence of g force. Human beings have experienced weightlessness for very brief periods in aircraft on zero-g trajectories, i.e., trajectories in which a lifting force equals or cancels the downward force of gravity. Weightlessness will be one of many unfamiliar environmental elements to be encountered by man in space. (See *g.*)

INDEX

MAJORITY LEADER JOHN W. McCORMACK, Chairman of the Select Committee on Astronautics and Space Exploration of Congress, describes this book as the most comprehensive study of space technology ever prepared in a form usable by laymen. The SPACE HANDBOOK was written by The RAND Corporation at the request of the Select Committee. It answers in nontechnical terms the principal questions related to all fields of space operations, civilian as well as military. It provides, according to Hanson W. Baldwin of the *New York Times,* background of great importance in assessing our present position and future course in the space age.

THE SPACE HANDBOOK CONTAINS CLEAR and comprehensive descriptions of the technical aspects of space flight, the uses to which astronautics can be put, and the current and foreseeable future of astronautics in this country and the U.S.S.R.

IN PREPARING THE SPACE HANDBOOK, the staff of The RAND Corporation was responding to a request from Congress to fill a need for better public understanding of this important new field of science. The RAND Corporation is a nonprofit research organization engaged primarily in research for the United States Air Force.